once upon a universe

ONCE UPON A UNIVERSE

Not-so-Grimm tales of cosmology

Robert Gilmore

C

COPERNICUS BOOKS
An Imprint of Springer-Verlag

© 2010 Springer-Verlag New York, Inc.

Published in the United States by Copernicus Books,
an imprint of Springer-Verlag New York, Inc.,
a member of BertelsmannSpringer Science+Business Media GmbH.

Copernicus Books
37 East 7th Street
New York, NY 10003
www.copernicusbooks.com

Book design and technical illustrations by Jordan Rosenblum

Library of Congress Cataloging-in-Publication Data
Gilmore, Robert
    Once upon a universe : not-so-Grimm tales of cosmology / Robert Gilmore.
        p. cm.
    Includes bibliographical references and index.

    1, Cosmology—Popular works.   I. Title.

QB982.G55 2003
523.1—dc22   2003062210

ISBN 978-1-4419-3059-0

Manufactured in the United States of America.

Printed on acid-free paper.

9 8 7 6 5 4 3 2 1

*I dedicate this book to my wife and family,
especially the growing new generation.*

acknowledgment

I wish to thank the Space Telescope Science Institute for permission to use images from the Hubble Space Telescope. These appear in the background of some of my illustrations. One in particular, the galaxy NGC 4414, appears in several pictures. It might be an interesting task for the reader to identify all the instances.

CONTENTS

• • • • • • • •

introduction

ix

the first tale

THE PRINCE AND p

(*A quest for the nature of motion*)

1

the second tale

SNOW WHITE AND THE PARTICULARLY LITTLE PEOPLE

(*Into the deep basement of our world*)

41

the third tale

ALI GORI AND THE CAVE OF NIGHT

(*Consider the heavens*)

77

the fourth tale

JACK AND THE STARSTALK

(*Spacetime and gravity*)

119

the fifth tale

WAKING BEAUTY

(*The big bang and after*)

157

the sixth tale

CINDERENDA AND THE DEATH OF STARS

(*The life and fate of stars*)

193

epilogue

(*A somewhat inconclusive conclusion*)

223

further reading

225

index

227

introduction

Throughout the ages people have been fascinated by the vision of the night sky. It conveys an impression of remoteness and of eternity. It is clear that we are looking beyond our Earth, but our understanding of what we see, of the Cosmos that lies out there, has varied considerably over the centuries.

At one time the Cosmos was seen as quite small. It was larger than the Earth, obviously, but it seemed clear that our good stable Earth must be at the center of the Universe, and that the Sun moved about us, together with various planetary wanderers in the sky. Around and enveloping all this celestial activity was the sphere of the fixed stars, which provided a sort of ornamental backdrop to the drama of our lives.

The notion that the Earth was at the center of the Universe was common in most early religious and philosophical views and was clearly required by our own proper self-esteem. Then came the time when Copernicus, Galileo, and others demolished our central position and sited the Sun and not the Earth at the center of things. Isaac Newton put the Sun-centered view on a firm basis with his theory of Universal Gravitation, which described how the planets moved in their paths.

Measurements of the distance to even the nearest stars had shown that the Universe was a very large place indeed. Newton consequently held that the Universe was not only large; it was infinite. He accepted the old belief that the heavenly region was eternal and unchanging, and this was now applied to the space filled with stars. It followed that the spread of stars must be infinite, as his theory showed that otherwise it would collapse toward its center under its own gravity. If space were infinite with the stars evenly distributed, then gravitational forces would balance, and there would be no center toward which it might collapse. Following Newton's lead, astronomers believed that the Cosmos should extend indefinitely into the past and future as well as into space in every direction. One problem with this picture is presented by what is known as Olbers' Paradox. Put most simply, this asks, "Why is the sky dark at night?" One is tempted to answer, "because the Sun is hidden by the bulk of the Earth," but the Sun is not at the heart of the question. If the Universe is infinite, and more or less uniformly populated with stars, then no matter in what direction you look, your line of sight will, sooner or later, end on a star. The entire night sky should look as bright as the surface of the Sun. If this were so, it would be very bad for us, but quite clearly it is not the case.

The picture of the Universe that led unavoidably to Olbers' Paradox was of course wrong. It assumed the Universe is eternal and unchanging, and that there has consequently been enough time for space to be filled by starlight from remote stars. In fact, the stars have finite lifetimes and, more importantly, the Universe itself has not lasted forever. It has had a long life by our standards, but not infinite.

What we observe is not the Universe as it actually *is* now. What we mean by "now" is a little difficult to specify because of the effects of relativity on time in a moving and expanding space, but apart from such complications there is a simple reason why we cannot see the Universe "now." That reason is the finite speed of light. We see things because light from them finds its way to our eyes, and it takes time for it to get there. We are *always* staring into the past. When we look at things in the same room as ourselves, we look back in time a mere few billionths of a second. When we see the Moon, we see it as it was about a second previously. When we look at stars, the time delay is much more significant. The distance to stars is measured in light years, and this is the number of years that it takes light from them to reach us. For distant galaxies it can take millions and even billions of years for their images to travel to us, and we see the most distant galaxies as they were in the early years of the Universe.

The Cosmos contains both things that are very small and things that are very large. We live somewhere in the middle; we are inconceivably larger than atoms; even more inconceivably smaller than galaxies. In our day-to-day life we are not especially aware of either end of this scale of being, though atoms, of course, do affect us since we are composed of them. We are aware of the Sun, since we rely on its light and in one way or another depend on it as a source of power. In the longer term we are indebted to the stars as the source of those very atoms that make up our bodies. On the very large scale we find the effects of the very small. Quantum effects and the physics of elementary particles play a central role in the Big Bang that began our Universe, and in the continuing life and death of stars.

It is unarguable that there is much we do not know. Philosophers sometimes speak to us of things that we cannot know, even in principle. In the nineteenth century it was stated authoritatively that we could never know the composition of the stars. In fact this is something we now do know. The light from the stars tells us what atoms and even molecules are present. Quantum physics shows us that light is emitted from an atom when electrons change between levels of definite energy, and that these levels are specific to that atom. The light comes as a sequence of precise frequencies, a line spectrum, and this is a unique fingerprint that identifies the atom it came from. The starlight can tell you even more. When the star is moving rapidly away from you, the lines in the spectrum are all shifted to lower frequencies. This *Doppler red shift* means that you can tell how fast a star is moving away from you.

The first observation that hinted at the Big Bang came when the light from distant galaxies was examined and it was found that, on the whole, the galaxies were moving away from us, and the farther away they were, the faster they receded. The galaxies are all moving away from one another, and because the ones farthest away are traveling proportionately faster, they must all have started out from much the same place at the same time. They have all come out of the Big Bang. In its earlier stages, in a fraction of a second so tiny that it is impossible for us to imagine, energy and density were so extreme as to be completely outside our experience. It was not long, though, much less than a millionth of a second, before the Universe reached a state in which the energy of matter, though still far from our normal experience, behaved in ways known to us from particle accelerator experiments that study the nature of matter at very high energies. Physicists have calculated how the Universe would be expected to have developed from that time onward. They find a steady development through a time of free particles, followed by the formation of nuclei and atoms and then a condensation of the matter into stars and galaxies. The calculations give numbers that are surprisingly in agreement with quantities that we can now measure. The distribution of light elements is as predicted. When atoms formed in the early Universe there was a great blaze of light that has since cooled and stretched to longer wavelengths. This is now seen as a background of microwave radiation that fills space, coming uniformly from all directions with a spectrum and energy that is in uncanny agreement with calculation.

The Universe is a big place and there is room in it for wonderful new discoveries aplenty. Yet what we already know is remarkable enough.

Robert Gilmore
August 2003

INTRODUCTION TO

the first tale

"Once upon a time there was no Universe," began the Story-teller. He paused and thought for a moment.

"No, that is not quite right," he corrected himself, "I cannot say 'once upon a time' because Time and Space began and developed together with the matter that is in the Universe. They were born with no father and no mother—and as for the midwife, who can say?"

His audience, who had been listening quite attentively, looked at one another. "I don't see how that can be so!" exclaimed Rachel. "There must have been time. There is always time (you could tell that she was quite young). How could anything happen if there is no time for it to happen in?"

"You have no reason to believe that anything was happening before the Universe first appeared with the dramatic entrance that we call the Big Bang. Everything that we know about, almost certainly everything that we can know about, has happened since the Big Bang. Subsequent events have most certainly required time in which to happen, but

time had then become available. The Big Bang provided the stage–the space and the time–as well as the cast of particles that make up the matter around us. The way these particles moved after they were formed has brought us eventually to where we are today.

"Because such movement is so vital I shall begin by telling you a story to illustrate the nature of motion."

"Why?" asked Adam. "Why don't you just tell us about the Big Bang itself?"

"That must come later," was the firm reply. "Since the Big Bang was the beginning of the whole Universe, it would make little sense to talk of it if you do not know what it is that was beginning. You would be hopelessly confused."

"So you are going to tell us what the Universe is like now," said Elizabeth, just to be sure about things.

"Not exactly," replied the Storyteller. "I cannot tell you what it is like now because there is no sensible way in which you can even talk about now for the whole Universe. Look at the stars." He looked intently at his audience as he said this, and such was his command of their imaginations, they

seemed to see the clouds part and the black night of space revealed to them. Its remote depths were sprinkled with a multitude of stars, stars everywhere.

"There you see a small sample of the Universe around us, but you must not believe that you are seeing it 'as it is now.' The light from those stars has taken a long time to reach you, and you are seeing them as they were years ago, thousands, even millions of years in the past. As you look outward you are seeing samples in time from a changing cosmos, and that change involves the nature of time and space and motion.

"Before you can usefully talk about the Universe you need to know about space and time and how both are tied up with motion in surprising ways. You need to consider what motion is, what is moving and what is at rest, and whether that is a question you can actually ask. For that reason our first story will be about motion and change, about how things move and how you see them move. It is about velocity and it is about energy and momentum, and so it becomes also a story about time and space themselves. It is the story of 'The Prince and p'."

"What does a pea have to do with motion?" asked Jordan in some confusion.

"I didn't say 'a pea'," replied the Storyteller rather irritably. "I said 'p'. You must listen more carefully. The symbol 'p' is used almost universally to denote momentum and that is what I meant. Now sit down and hear the story." He cleared his throat, adjusted to a more comfortable position on his seat and began his tale.

"A certain king had three sons and as was the custom in that kingdom, as each son came of age he called him into his presence and assigned him a quest. In due course he called his youngest son. . . .

The
PRINCE
and p

(*A quest for the nature of motion*)

Prince Kevin came quietly into his father's throne room and stood respectfully before him. The king was standing, looking out of the window with his back turned to the lad, but after Kevin coughed gently the king turned around to face him.

"Ahem! Ah yes, m'boy. It's time now, you know. Time you went out on your special quest, the one you must accomplish before you come into your inheritance." With these introductory words the prince's father formally addressed his son: "Anyway, as your father and your king, I now command that you go out into the world on a quest to discover the nature of motion."

"WHAT?" exclaimed Kevin, noticeably less respectfully than before. "What sort of a daft quest is that? What has become of killing dragons and searching for magic rings and all that sort of thing? You know, the traditional sort of quest."

The king looked unhappy. "Well now," he said, "you know that I sent your eldest brother George out to kill a dragon that had been troubling the next country. A lovely boy, your brother George," he said reminiscently. "Brave as a lion and not an ounce of guile in him. Straight up to the dragon he went and challenged it to a fight. Wonderful lad."

"The dragon ate him, didn't it father?" said Kevin.

"Yes, I am afraid so," sighed the king. "Anyhow, your second brother Casper was quite different. He was a much more subtle character. Intelligent he was, sharp as a needle. Some people even called him a bit devious. I sent him out to seek the magic ring of the Witch of the Western Wastelands."

"And did he succeed father?" asked Kevin. He already knew the answer, but he wanted to show that he was listening.

"Oh, yes. At least so I understand. He managed to distract and capture her by some means or other...."

"He imprisoned her for a hundred years in the heart of a giant oak tree, didn't he father?"

"Yes, something like that, something ingenious but a bit underhanded. Anyway, he got the ring. He had the witch's magic ring that could command the powers of Earth, Water, Air, and Fire and would give him access to all the riches of the world. He had thus fulfilled his quest and was entitled to come back and claim my kingdom." The king turned and looked out at his country, at the two valleys of rather uncertain fertility and the surrounding dark, impenetrable forest.

"He never came back, did he father?" said Kevin gently.

The king sighed even more deeply than before. "No my boy, he never did. That is why I am sending you on a quest that should not be too dangerous if you fail or too tempting if you succeed. So off you go and be as quick as you can about it."

Thus dismissed, Kevin left his father's presence, put on his traveling clothes and set off. It took little time to cross the fields near the castle and not too much effort to enter the seemingly impenetrable forest and walk through its depths, which were reputed to be trackless. After a little time spent in forcing his way through the undergrowth he found himself following what could be described as a faint trail. This led him on a winding route between the massive trees. The trail grew more definite with each step. It was no longer just a grassy path through the undergrowth but had developed a hard rocky surface. Kevin rounded another bend and saw ahead of him a straight section of the trail that rose up a long incline. At the top of the slope stood a tall man with a black gown hunched around his shoulders. His angular features and the way his head thrust forward gave him the look of a bird of prey. He watched quizzically as the young prince climbed laboriously up the hill toward him.

"Good day," said Kevin politely. He had been taught to be polite to his elders and indeed most other people. "May I ask who you are, please?"

"I am frequently known as the Dominie, boy. You may, if you prefer, refer to me as the Pedagogue or, if you really must, simply as the Teacher. On the whole I prefer the Dominie. When you address me, you may just call me Sir. It is my duty and pleasure to impart knowledge to all who seek it." Kevin was delighted. It looked as if his quest might be over rather quickly. Perhaps he could even get back to the castle in time for lunch.

"I am on a quest for the true nature of motion," he said brightly. "Would you be able to help me, please?"

"Oh yes, I can help you, or at least I can help you to begin your learning, or at least to learn how to begin. To truly learn you must teach yourself. Tell me young man, what do you understand by motion?" he asked searchingly.

"Well, it is moving from place to place, of course."

"Fair enough, I suppose, if a bit trite," replied The Dominie grudgingly. "But you can do better than that. Are we presently in motion, would you say?"

"No," replied Kevin fairly confidently. "I was in motion, of course, as I approached you, but now we are both stationary."

"Are you sure about that? How can you say so confidently that you are not moving?" The tall figure opposite fixed him with a challenging gaze.

The prince was a little irritated by the question, though at the same time he had to admit that he was a bit daunted by the questioner. He answered boldly, "I can tell well enough when I am walking and when I am not–and just look around you! Look at all the trees and rocks, look at the ground we stand on. Nothing is getting any closer or farther away. We are clearly motionless."

"So you take your opinion from the company in which you find yourself." Kevin somehow felt reprimanded as his companion spoke to him. "The trees around you are keeping their distance from one another and so you maintain that not one of you is moving. This is very democratic no doubt, but not perhaps very good sense. Come with me," he commanded suddenly.

The Dominie turned abruptly and strode into the undergrowth. Kevin followed him and shortly thereafter they came to the
bank of a wide river, flowing deep and swift in front of them. The
Dominie continued to stride confidently ahead and stepped, apparently without looking, right off the bank and onto the stern of a
long punt-like boat that was drifting past. Kevin was about to
follow his guide and step off confidently, but something made him
look down at the last minute. He stopped abruptly with his arms
flailing for balance as he saw that the boat had drifted on a little
farther and he was about to step confidently into the river. He
turned and ran along the bank after the boat, finally managing to
catch up with it and board with an inelegant scramble. He landed
on the decking beside his mentor, who ignored his antics and went
on talking as if Kevin had been there all the time. The two walked
side by side toward the front of the boat.

"Now you see the trees and rocks as moving steadily past us.
Does this make you think they are no longer at rest?"

"Not at all!" responded Kevin, still convinced that he had as
good a right to his opinion as anyone. "The trees have not moved. It
is we who are moving and so naturally we see the trees move past."

His companion did not answer directly and they walked on until
they reached the bow of the boat (it did seem a remarkably long

boat). He turned and they began to walk back toward the stern. They walked at such a pace as to match the drift of the boat and so held their position beside a particularly large tree on the riverbank.

"Would you say we are in motion now?" asked The Dominie abruptly. "You are now definitely walking along are you not? I can see your legs moving," he said bitingly.

"Yes," admitted Kevin stubbornly, "but my motion is balanced by the motion of the boat. Overall we are not moving at all. You can tell because that tree, which is fixed to the Earth, does not appear to be moving. It is not moving and so we are not moving either." That seemed pretty conclusive to him.

"So once again you argue that might is right and that the Earth and the trees on it must be at rest simply because it is large," was the scornful reply. "Come with me!"

Abruptly he strode off the boat and onto the riverbank. Kevin followed with confidence this time, as the riverbank was so long that he could hardly miss it. As they started along the path by the river, Kevin heard a splash from the water nearby.

"When we were on the boat, you argued that its velocity had combined with ours as we walked along it in such a way as to result in no motion at all. In that case the velocities were all along the same line. Here you may see a rather more interesting case of combining motions, one where the movements are in quite different directions, as is usually the case," remarked the Dominie. The prince looked to see what he meant and spotted a water rat swimming gamely across the water. Its objective was a point directly across the river, but being rather naïve, it was attempting to swim straight across. Unfortunately the river was wide and flowing quite fast and the current carried the rat downstream, so that it would arrive at the opposite bank well below the intended destination.

"You see that the flow of the river and the rat's own speed through the water have combined. It is as if the velocities were two sides of a triangle and the result is an overall movement of a size and direction given by the third side."

. . . .

Addition of velocities

When the motion of an object combines two motions, like a swimmer carried along by a current, the overall motion is given by *triangular addition* as shown The overall velocity will differ from the separate velocities of the swimmer and the current, and the velocity relative to the land is different from that relative to the water.

This addition rule works for ordinary velocities, those much less than the velocity of light. Close to the velocity of light things start to get a bit funny.

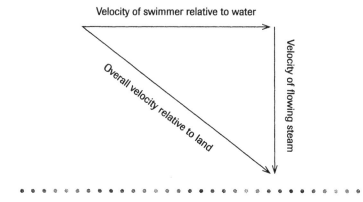

Eventually, as they watched they saw the dispirited rat reach the other bank and climb dripping out of the water. With a sigh of resignation it began to plod slowly upriver, moving along the third side of the triangle formed by its intended path and its actual motion. They could faintly hear it grumbling resignedly as it went.

Turning away from the disconsolate rodent, they walked briskly along the path and soon came to a small clearing in which there was, rather surprisingly, a large blackboard of the type that Kevin remembered from his schooldays. This board was *very* black and there was a hint in the blackness of distant stars scattered over its surface. From somewhere within his gown the Dominie produced a piece of chalk and drew a large circle in the center of the board. He stepped back and the young prince saw the circle fill with a blue color overlaid with white cloud-like streaks and a white area in the center. The shape was rotating around this central point.

"That is the Earth, seen from one of the poles. Consider two people on opposite sides of the Earth." A quick motion with the chalk and two tiny stick figures appeared. Both figures began to wave furiously to the camera.

"As the Earth rotates, you may observe that these two are moving in opposite directions. Which, if either, is motionless? They cannot both be. Furthermore, the Earth does not only rotate about its axis. It also moves in an orbit around the Sun."

The image of the Earth, still bearing its little figures around as it rotated, moved across the board and off one side, leaving only a sprinkling of stars on the otherwise black surface.

"In truth you cannot say that anything is absolutely at rest. Motion is relative. All motion is relative!" he said didactically. "The Universe is a large place and everything in it is moving relative to most of the other things in it. There is no fixed point and all is in turmoil and motion.

• •

Absolute motion

There appears to be no such thing as absolute motion. Everything in the Universe is moving relative to hordes of other things and you may only say how fast one thing is moving relative to another.

• •

"If all motion is relative you cannot be sure whether anything has a velocity in and of itself. What, you might ask, is the essence, the discernible measure of motion that any object may seem to possess?" Kevin thought that was what he had originally asked. Maybe he was now about to learn the answer and could go home.

As he was contemplating this possibility, a butterfly came fluttering along the path and collided briefly with his forehead. It produced no discernible effect (apart, perhaps, from a hurricane in Florida[1]) and he walked on scarcely noticing. Almost immediately afterward a portly figure came panting along the path, waving a large butterfly net. He was so intent on his elusive prey that he ran straight into the slight figure of the prince, who was knocked head over heels into the undergrowth by the side of the path. As he sat up staring at the ponderous individual who was still thundering along, without appearing to have deviated from his path, he heard the dry voice of the Dominie from behind his right ear.

"That incident demonstrates a significant aspect of motion, the property of *momentum*. It shows an effect of the *inertia* possessed by bodies. Both of your assailants were moving at the same speed, but you will have noted that the motion of one of them had much a

• • • • • • • • • •

1 It has become a cliché that the flapping of a butterfly in a jungle in Borneo may cause a hurricane in Florida. It would not, of course, *cause* a hurricane. The main causes are the buildup of pressure and water vapor in the atmosphere. The point made is that the actual formation of a hurricane depends upon many factors, including perhaps the butterfly but also innumerable other things that might happen around the world.

greater impact. It was because of a
difference in their masses. They
may have had the same velocity
but they had different momenta
because one was much the heavier. Momentum is defined as velocity multiplied by mass and, as you
have just seen, it is momentum
that matters in a collision. Much of
what goes on in the Universe
depends on collisions of one sort
or another, usually just collisions
between tiny particles but occa-
sionally larger things like planets or galaxies. In all cases momen-
tum is critical. Momentum is of such importance in consideration
of motion that it is allocated its own special symbol to represent it:
momentum is invariably known as 'p.'"

"Why?" asked the prince. The Dominie was temporarily thrown
off his lecturing stride. "It just is!" he snapped, then continued in
more measured tones. "The important feature of momentum is that
it is conserved in collisions, the total momentum remains the same
after a collision as it was before. The object that has a lot of momentum will usually share some of it with anything else involved. The
gentleman who collided with you had so much momentum because
of his considerable mass that he was able to give you some of it
without much apparent loss to his own motion."

Kevin continues to stare at the bulky figure running on down
the path and, just as he was about to look away, another similar
character came thundering round the corner and the two collided
with dramatic effect. Both were brought to a halt, then bounced off
one another and sat down abruptly. After a moment they scrambled laboriously to their feet, exchanged a few words, and then
headed off together in pursuit of the butterfly.

The prince's companion took this opportunity to remark "After
the comments you made about addition and subtraction of velocities while we were walking on the boat, it is worth noting that
momentum, which is after all velocity multiplied by mass, will add
and subtract in the same way as velocity itself. Two equal and
opposite velocities will cancel one another, as you noted aboard
the drifting boat. In the same way equal and opposite momenta
will cancel in a collision, as you have just seen."

Before they could resume their walk, they were interrupted
again, this time by a roughly clad person who sprang out in front

of them, whirling around his head a staff to which was attached a spiked ball on the end of a long chain.

"Stop right there, don't move an inch. Hand over all your money," he roared rather illogically.

To the surprise of the prince, and indeed of the prospective robber, the Dominie produced from beneath his gown a long pointer of the type often used by speakers during illustrated talks, and this he held immovably in front of him. Immovably was the proper word, for it did not even waver when the outlaw's weapon crashed into it. The chain hit the rod and wrapped itself around it. As the length of chain wound shorter and shorter, Kevin noticed that the heavy ball moved ever faster until it finally collided with the staff with an abrupt thud. The thief looked at it in disbelief, and during his momentary shock Kevin leapt upon him, overpowered him, bound him with several items of his own unsavory clothing, and then pushed him into the undergrowth. The weapon he threw far into the woods, producing after a moment a loud splash and a brief burst of grumbling–something about a spiked metal ball that suddenly appeared from nowhere and carried someone's butterfly net into the something river.

"There, you see that not only is the pen mightier than the sword but also the blackboard pointer is greater than the morning star.[2] This encounter has served to give you an illustration of the conservation of *angular* momentum. This relates to anything that is swinging around some central point. The measure of this quantity is the mass of the object multiplied by its velocity, as for normal momentum, but now multiplied also by its distance from the center of rotation. As the metal sphere came ever closer to the rod around which it was rotating, this distance decreased and so, since its mass did not change, its velocity became greater."

• •

Angular momentum

The conservation of angular momentum is clearly evident in astronomy. When something rotates around a point it has *angular momentum* defined as

$$\mathbf{J} = m\,\mathbf{v} \times \mathbf{r}$$

• • • • • • • • •

2 The medieval weapon that consisted of a spiked weight on the end of a chain was known rather charmingly as a morning star.

Here m is the mass of the object concerned, **v** is its velocity, and **r** its distance to that central point. In the absence of other influences, angular momentum remains constant, so a planet in its elliptical orbit around the Sun moves more slowly when far from the Sun, faster when it is close. This leads to *Kepler's second law* of planetary motion, that the line from the Sun to a planet sweeps out equal areas in equal times. The shaded areas in the diagram indicate areas swept out in equal times by the line from the Sun to the planet. Near the Sun the planet is travelling more quickly, so the planet sweeps out a wider region.

There is at all times a gravitational force pulling the planet toward the Sun, but as it is directed toward the Sun it does not affect the planets angular momentum *around* the Sun.

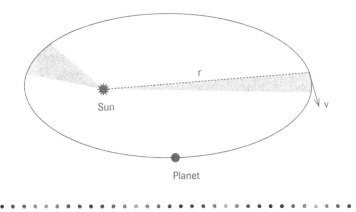

• •

"But then it suddenly stopped!" protested Kevin. "It does not look to me as if any measure of motion could be constant then."

"But in point of fact it was, if you take the wide view as you must. Momentum *was* conserved. When the ball hit my pointer, then its momentum was transferred to that and, since I held it firmly, was transferred through me to the Earth beneath. We have seen that the sheer size of the Earth may not be a good reason to say that it is not moving, but its mass is great enough to absorb a little momentum without any apparent effect. In fact, when you begin to run along the ground the momentum that you build up will be balanced by an opposite momentum for the Earth itself, but such is the mass of the Earth that you would be hard pressed to discern this.

"Now I must leave you," he said abruptly. " You may continue your education in the nature of motion at the hostelry just down the road. I hope you benefit from the opportunity." He turned away, after pausing briefly to scribble on the bottom of the prince's cloak:

Could do better, room for improvement. C–

Kevin continued on down the path and shortly turned a corner and saw in front of him a friendly looking tavern. In front swung a sign that said
A board by the door stated:

THE TEACH INN
ROOMS for improvement
FOOD for thought
DRINK deep of the spring of knowledge

This was obviously the place he had been told about, so he walked up to the door and entered. Inside he found a dark barroom with rough tables scattered around, most of them unoccupied. A long wooden bar extended across one end of the room and near it were stacked a number of barrels. Kevin noted that instead of the customary XXXX the nearest was marked with X^4. Behind the bar stood the innkeeper, a smile of welcome spread across his homely features.

"Good morning, young sir!" he greeted the prince. "Have you come to partake of sustenance for your mind? What is your pleasure? What may I offer you?"

"What do you have?" retorted Kevin rather shortly. This did not seem quite like the bars near his father's castle, and he was not at all sure what he might reasonably ask for.

"Well sir, we have all sorts here, from simple arithmetical brews to some long matured elixirs of the physical sciences. For some, an undemanding draft is more appropriate." He indicated a rough looking group who were huddled round one of the tables. There was much discussion and a lot of sniggering. "They have come in for their daily intake of straightforward arithmetic, but they are still stuck with vulgar fractions.

"For a gentleman of your quality," he went on ingratiatingly, "who is evidently on some variety of quest as gentlemen of your age tend to be, I would venture to suggest a little distilled energy. There is plenty of choice." He turned to indicate the array of bottles on the shelves behind him. "A supply of energy should set you up wonderfully for all the exertions and dangers that will lie in your path."

Kevin rather wished that the innkeeper had not mentioned possible dangers, but he still scanned the bottles with interest. There was certainly an enormous variety of them, in all shapes

and sizes and bearing a bewildering array of labels. The innkeeper indicated various possibilities in turn. One carried a symbol like a stroke of lightning, "That is electrical energy, of course". Another bottle bore a strange pattern of swirling lines, "That one is magnetic energy." One label showed an eye, "Light energy," another a picture of an ear, "Acoustic energy." There was one squat, bottle with a rather disconcerting picture of a mushroom shaped cloud, "That is nuclear energy, I am not quite sure that we are licensed to serve that one." A particularly large bottle that sat on top of the bar had a label that showed a lot of little arrows pointing in all directions, "That is heat energy, we pour the dregs from all the other bottles into that one," muttered the innkeeper.

A bottle caught the prince's eye that had a picture of a large determined arrow pointing firmly in one direction on its front. When he asked about this one the innkeeper replied, "That is kinetic energy, probably the most obvious and in a way unsubtle of our various brands. It is the energy of motion and is associated with any object that moves. Kinetic energy and momentum are in a way part of the same blend.

"If you are to be moving about you will certainly need kinetic energy, the energy of motion. This is the undirected but enduring companion of momentum, of which you have already heard tell. Where momentum is mass times velocity, Kinetic energy is one half mass times velocity times velocity again. At least it is by the reckoning of we humble folk. The folk in the *fast set* might tell you different," he added with a glance toward a dark corner of the room. "It might be more convenient though if you were to carry your energy in some other form."

Another label showed a rather attractive mountain peak, "That is potential energy. It is energy that is not obviously present but can become so, in the way that a rock may roll down a mountain, and as it goes it will gather kinetic energy and indeed momentum. This kinetic energy is provided by a decrease in its gravitational potential energy. You see one form of energy may convert to another, as is its custom.

"There does seem to be plenty of variety," offered the young prince, "though I am not familiar with any of them. What would you advise?"

"Well, Sir, it doesn't really matter very much what you choose." He leaned confidentially over the bar. "Actually they are all essentially the same. There is basically the same stuff in each bottle. Here try some of this" He poured something from one of the bottles into a tankard and held it out to the prince, who sniffed at it cautiously. As far as he could tell it was just alcohol.

• •

Energy and momentum

A moving object of mass m and velocity v has the properties of kinetic energy ($\frac{1}{2}mv^2$) and momentum ($m\mathbf{v}$).

Energy and momentum are useful because they are conserved quantities. That is to say, they have the same total value after a collision or other interaction as they had before.

Momentum is a *vector quantity*. Its value depends on the direction of motion, and objects moving in opposite direction have opposite momentum. Two particles with equal and opposite momentum have a total momentum of zero, so after a collision they may both end up stationary even though momentum is conserved.

Energy also is conserved, though it may convert from one form to another. If a ball runs up a slope it will come to rest as its kinetic energy converts to gravitational potential energy.

• •

"There, sir, you have chemical energy. That is energy locked up in the bonds of chemical compounds from which it may be released, often by some form of burning. This may happen slowly, as when the food that you eat gives you energy to move about, or it may happen rather quickly, as when a charge of gunpowder ignites to propel a cannon ball to a great height." There was a loud whooshing sound and the tankard soared into the air, rising to a height of about three feet above the bar.

"Now you have potential energy, gravitational potential energy. Because an object is higher in the Earth's gravitational field, it may fall." After seeming to hang motionless at the top of its trajectory for a fraction of a second, the tankard did fall, faster and faster. "And now its potential energy is being transferred back to kinetic energy as it falls. This will continue until…" The tankard landed on the bar with a loud echoing thud ("Acoustic energy"). The prince reached out to grasp it and found that the tankard was quite hot to the touch–"There you have thermal energy. Most energy ends up as heat eventually, the random distribution of motion among the atoms of any material.

"But now," he continued, "you will be needing a good supply of energy to sustain you on your quest. You just take this with you, young sir, and it should see you through." So saying the landlord produced a large flask that he handed to Kevin, who took it with absent-minded thanks and put it in the pouch he carried by his side. His attention was distracted by a tankard in the hand of a neighbor at the bar. It seemed to be full of flashes of lighting as if it contained a miniature thunderstorm, "Electrical energy," explained the innkeeper, noticing the direction of his gaze. "Some of it converting to light and acoustic energy," he added as the miniature storm flashed and rumbled.

"And what sort of quest did you say you were on?" he continued with a display of professional interest as he wiped the surface of the bar. Kevin could not remember having said anything at all about a quest, though his host seemed convinced that he was on one. As it so happened that he was on a quest, he replied that he was searching for the nature of motion.

"Ah, well now, if that is the case you will be wanting to go where the real movers are. You must mix with the fast set. We common folk hardly move at all by comparison, so you must make the acquaintance of the aristocracy of action, the faerie folk of furious velocity."

"And how would I do that?" inquired the young prince.

"Well, to be sure you are in luck, for there is one of them gracing my humble inn at this very moment. Your honor," he raised his voice, "may I introduce to you Prince Kevin, who is on a quest for the true nature of motion."

Kevin could not recall having mentioned his name, but nonetheless turned to look toward the dark corner that the innkeeper was now facing. He saw a tall, dark shape arise from a solitary table and move out into the light. He was revealed as a figure of more than normal height, dressed in fantastically ornamented silver armor. From behind the intricate metalwork of his helmet, bright eyes regarded Kevin.

"A prince, eh? You must come and join us. A young man of your rank should not be consorting with commoners and nobodies. Allow me to introduce myself. I am an Earl of Faerie; Ingus is my name. Come and ride with me and with my companions. It will be a wild ride," he warned.

"Aye, that it will," agreed the innkeeper. "It is a rare sight should you happen to see any of the faerie folk as they arrive at my inn. They ride like the wind."

"Not so," corrected the Earl fiercely. "Never do we ride as slowly as anything like a mere wind. We ride," he claimed proudly, "more like a moonbeam."

"But moonbeams don't move," protested Kevin.

"Sure and they do," responded the Earl. "It is just that they move too fast for the common gaze to detect. They do move, though. All light moves and always it moves at the same speed: the speed of light. We folk of Faerie are not accustomed to dawdle, and so in our lunatic gallops through the woods at night we travel almost as fast as moonbeams themselves. There is in this Universe of yours a Limiting Velocity and this is the *velocity of light,* so we ride as close to this speed as we may. Come," he finally commanded.

. . . .

Relativity and the speed of light

Einstein's theory of special relativity was in fact a departure from the notion that all motion is relative. It is a remarkable observation that, no matter how an observer may move, he or she sees the speed of light as being exactly the same and never sees anything moving faster than this. From this single fact Einstein predicted the strange distortions of space and time that comprise his theory.

. . . .

The prince followed him out to find, standing by the inn door, a great white horse that was itself the color of moonbeams. Its eyes shone with a silver-white inner blaze. Beside this horse, and almost hidden by its bulk, stood a small pony held by a sprite in a green jerkin and tights. "This is Puck," said the Earl simply.

"Oh," inquired the prince, "are you the same Puck as is featured by Shakespeare in *A Midsummer Night's Dream?*

"Sure and I am not! What would I be doing with that English fella?" he replied proudly. "No, I am the Puck of the Irish."

"Come now," said the Earl, "let us mount and away. We must haste to show you what real motion can be. We will gallop with such haste that scarcely could even a moonbeam catch us. You will ride behind me on my steed," he added, turning to the prince.

Kevin was not so sure that he felt like doing this. The horse looked even more enormous the closer he came to it, and there was a certain wild abandon about the Earl that he did not quite trust. Puck noticed his hesitation and, coming up to the prince's side, hastened to reassure him from somewhere in the vicinity of his elbow.

"Ach now, me darling, you need not worry at all, at all. You will be perfectly safe if you travel with Earl Ingus, so you will. It should be quite an exciting journey for you, traveling close to the speed of light and all that. You should not be anxious though; we do it all the time. I can put a girdle round the Earth in a hundred and fifty milliseconds, so I can."

The Earl vaulted into his saddle, his strange armor glinting brightly now that he was out of the dim interior. Although he still had lingering doubts, Kevin climbed up behind him and their ride began. Certainly they moved fast, but it did not seem quite so extraordinarily fast as he might have expected. He had, after all, been told that he would be moving almost as fast as it was possible for anything to move–the limiting velocity c. He began to look more carefully at the woods as they flashed by and noticed that they were strangely still – unbelievably still. A little waterfall that cascaded down a rock face was rigid as if carved from ice. A swallow hung motionless in mid air, frozen in the act of pursuing a cloud of insects that were motionless also, as if mounted on invisible pins on a specimen board. He remarked on this out loud and was answered by Puck, who was keeping pace on his little pony with no apparent trouble.

"Well now, you are right enough, to be sure, in what you say and aren't there after being two reasons for it," he replied. "For one, if you are to travel with us and if you are to be seeing anything that is happening, what with moving close to the speed of light and all, you must needs live a little bit faster than usual, so you must. Over the sort of distances you meet on Earth, the time that light takes to travel from one place to another is far too short for you to notice, so it is. Sure and it would be different if we were to look at the distances between stars and galaxies, as there even light may take a very long time. Here, though, you must think and speak, not by the second, but by the millionth of a second, if not faster.

> *If you can fill the unforgiving second,*
> *With a million microseconds worth of duty run,*
> *You'll find the time for every task that beckoned,*
> *Who knows? That may include a little fun.*

He quoted gaily.

"The second thing that you are after seeing," he went on more seriously, "is not indeed a thing that is within yourself. It is a thing of space and time: a twisting of the underlying structure of the Universe, so it is. It is called the Time Dilation effect.

"When you travel fast enough, you see things a bit differently. Actually you see them as they always are, but traveling close to the speed of light makes it that much more obvious to be sure. One of the things you see is that everything you pass slows down. Clocks move more slowly, people move more slowly, *time* passes more slowly. This is not after being an illusion of some sort, not at all. Time really is different, depending on how you look at it, or at least how fast you are moving as you look at it. Moving clocks are slow, do you see."

"But weren't you saying that we were moving past the clocks, not the clocks themselves moving?" said Kevin.

"Well now, doesn't it after be coming to the same thing? Sure, and all motion is relative, so there is no need at all to be arguing about who it is that is moving and who not. Each sees the other as moving and each sees any clocks carried by the other to be running slowly."

"But who is right?" asked Kevin. "Which clocks really are running more slowly?"

"We will have no preacher's talk of right and wrong," stated Puck firmly, guiding his pony even nearer to Kevin's right leg. "Sure and everybody sees what he sees and there's an end to it. We speak of each one being in a moving *frame*, a sort of set of yard-sticks with which each might measure distances round about it as it moves. Each frame must have not only yardsticks but also clocks, to be sure, since each has its own time as it moves along. Even if everyone is moving differently, none have preference. Different folk will see their neighbors as having different speeds, but each and every one will see light to be moving with the selfsame velocity. With lesser movement, someone who rushes toward your direction of movement will appear to you to travel faster, one who follows will appear slower. A stern chase is a long chase as they say at sea. For light, or rather for the special velocity of light, it is different. Space and time must in some way distort themselves to ensue that everyone sees this speed to be the same; for sure enough that is what they *do* see.

"Every observer will observe the times of other folks' frames as moving more slowly than his or her own. I should hope that you are not thinking that there is one absolute, imperialist sort of time that all must share. We will not be having anything of that sort here!" he exclaimed indignantly.

"There is no true and absolute time any more than there is one true and absolute position. People used to think they had a special place in the centre of the Universe, then they thought they had the only true time. Sure and they were wrong in each case. To experience someone's position and time you have to *be* there. There can be no location without representation. Yardsticks and clocks, space and time–they are all bound up together in a sort of composite SpaceTime," he concluded more calmly, but still firmly.

Kevin was rather taken aback by Puck's fierceness on this point and felt he needed a little time to chew on what he had been told. He crouched down behind the armored back of the Earl and took a morsel of bread from his pouch to give himself something else to chew on while he thought. The bread did not look very appetizing however, so he casually threw it away toward the trees. "NO!" cried Puck frantically as he saw this, but it was too late. The morsel of bread went sailing off at an angle toward the dense foliage.

When the piece of bread hit the nearest tree in its path the prince was astonished and horrified to see the broad trunk explode into fragments and the lofty crown topple forward as if felled by a woodman's axe. Scarcely had he seen one tree suffer this disaster than the tree behind it and then the one beyond suffered a similar fate. Further yet along the line where the bread had vanished from his sight, there continued vast commotion within the forest. One after another, great trees crashed to the ground in sequence like a line of giant leafy dominoes pierced by an arrow from a crossbow. The disturbance passed quickly from sight as the dense woods hid more distant tree falls. Kevin sat openmouthed, unable to believe what he had seen.

A short while later there was a brilliant blast of light in the distance and a sinister cloud rose into the air above the woods.

"What was that?" quavered the young prince. "What have I done?"

"Well now," replied Puck, looking slightly shaken himself, "as we are riding so close to the velocity of light, why anything you throw away will be after having much the same speed. Have you not considered what energy it must be having to be traveling so fast?"

Kevin admitted that he had not.

"Well now, you know that when something moves it must needs be having momentum and kinetic energy, do you not?"

The prince admitted that he had already been told of this.

"And when that something is moving close to the speed of light itself, then that momentum and energy become rather high, so they do. Do not be after thinking that, just because the *speed* cannot rise above the limiting value, the momentum and energy are so limited. Not so. The actual speed of motion through space may never rise above this limit, but the momentum and the energy may both rise *without* limit. Sure and this is just the way that space and time are behaving.

"Momentum and energy both depend on mass as well as on speed and they behave as if the mass were rising as the speed approaches that of light. The closer the speed to that of light the more steeply does the mass appear to rise. To be sure, you cannot prove the mass itself is rising. At such high speeds you cannot measure mass as distinct from energy, but the momentum and energy increase *as if* the mass were rising, and so we say that it is. Indeed we say that mass and energy are after being the same thing entirely.[3] At low speed the mass hardly changes at all, at all. When something is not moving, then it has no momentum, as you might expect, but it still has energy, a lot of energy. It has *rest mass* energy.

• '

Energy and momentum at high and low speed

Objects may not travel faster than the speed of light, but as they approach that speed their energy and momentum rise without limit.

At low speed momentum approaches zero, but energy does not. It tends to a finite limit that is given by $E = mc^2$.

This is called the *rest mass energy*.

In effect mass and energy are the same thing.

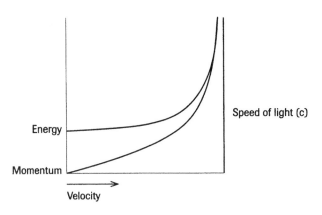

• •

Mass is energy and energy is mass, do you see. What the common people call kinetic energy is no more than the tiny *increase* in this mass or energy when an object is moving slowly. The smaller the rest mass that something has, the less energy you have to give it to make it go fast, but however small the mass it would still need an infinite amount of energy to make it go at the velocity of light. That is why it can't. Only something with no rest mass at all may travel at this speed. Even yon little scrap of bread you dropped was traveling so fast that it had a *lot* of energy, so from now on do not discard *anything.*"

Kevin said in a small voice that he was sorry and then held himself as still as he could for the rest of their journey. Finally, the Earl and Puck drew in their mounts to a trot and then down to a slow walk. They had arrived.

In the distance before them they could see a lofty castle with tall towers topped by conical roofs. It looked, not unreasonably, like a fairy-tale castle. Before the castle was an area of level ground, now a scene of bustling activity. As they rode slowly along they passed a section of field where a group of archers were firing at a target. The prince saw arrow after arrow speed swiftly to its destination. Though clearly moving pretty fast, each arrow took some time in flight. He wondered how quickly they were really moving and asked aloud if he was still living at the same rapid rate as on his ride through the forest.

"Yes, nothing has changed, you are still keeping fast company," he was answered.

"In that case," he said, "surely the arrows must have energy at least as great as the bit of bread I dropped. Why is there not a problem when they strike the target?"

"It is simply a matter of good design, so it is. Our best artificers cunningly contrive the targets so that they may absorb the energy and momentum of the missiles without harm. It would be a different matter were one of the archers to fail in his aim and miss the target, but our faerie archers do not do that," Puck answered rather smugly.

This was good news, as their path took them past the back of the target. Kevin saw a distant archer loose an arrow toward them.

• • • • • • • • •

3 The relation between mass and energy is given by the famous Einstein equation $E=mc^2$. As c is a constant, this really says that mass and energy are the same thing.

   At low speeds momentum is given by $p=mv$ and kinetic energy by $E=\frac{1}{2}mv^2$. Since v will be very much less than c, this is but a tiny increase in the rest energy mc^2.

This time, instead of the expected delay, there was an almost immediate "thunk" sound and he saw that same arrow jutting from the front of the target.

"How is that?" he exclaimed. "A few moments ago I could see that the arrows took some time to reach the target, but this one has arrived almost immediately after it was fired. What has changed?"

"Only your position," was Puck's answer. "I hope you are not after thinking that you actually see *what is happening.* All that you see is the light that reaches your eyes, and the arrows are traveling almost as fast as the light, so they are. When you were watching the flying arrows from the side, then the light from them took much the same time to reach you from each and every point along their path. You saw them as moving at much the speed as in truth they were. When you are behind the target, then the light and the arrows are both coming toward you and the light that set out when the archer fired off the arrow arrives but a little before the arrow itself. Thus it seems to you as if the arrow came in next to no time, since it arrived so soon after that same light which informed you of its release. As is so often the case, you cannot take at face value everything that you see.

"Come now. As you are a prince and hence noble, it becomes you to take part in a little jousting." They had arrived at a cleared area arrayed as for a tournament. Kevin saw two knights riding furiously toward one another, each with his lance held firmly in front of him. He noticed that their armor and trappings were of different colors. Those of the approaching knight were of a clear blue, while the knight moving away from them was accoutred in armor of an angry red. Although the area through which they were riding was fenced off and had obviously been cleared for the occasion, it was not entirely obstacle free. There were a number of deep craters in the ground, and the knights had to swerve carefully around these in their furious advance. To cheers from the crowd, the knights drew near and then struck one another. There was a brilliant flash as their energy was released and a great cloud of smoke and dust rose in the air.

Afterward there was another crater in the ground.

"Come along then, now is after being your chance." Before Kevin could think of a compelling reason to avoid the contest that would not make him seem unduly timid, he found that the Earl had hoisted him with little apparent effort onto an enormous fiery horse, placed a spear in his uncertain hand, and that he was galloping down the field of tournament. Just as he was wishing that

he had given even an excuse that
would have made him seem unduly
timid, he looked up and saw a blue
knight thundering toward him at a
quite unexpected speed, lance stretching out in front. Indeed he saw that
the lance *was* stretching out in front.
As the knight came nearer, his lance
grew and grew in a most unreasonable fashion, seeming to leap magi-
cally toward Kevin's heart of its own

accord. The lad jerked in panic and his horse swerved so that his
opponent missed his aim and went thundering past. The young
prince breathed a deep sigh of relief and struggled to try and slow
his mount. He turned in his saddle to look at his departing adver-
sary and now saw him transformed from a blue to a dull red figure
beginning to rein back his horse. He appeared to be moving much
more slowly than on his approach, while the great probing lance
had contracted to a short stunted spike. As they both reined to a
halt, the red tone of his opponent's armor faded to a more neutral
color. At last, Kevin managed to bring his mount to a halt and tumbled from the saddle to lie panting on the ground.

"There now, wasn't that exciting?" asked Puck as he helped the
young man to his feet. "Sure and it's experiences like this that
make the heart leap within your chest do they not?" Kevin thought
"stop" might be a better word than "leap," but there was obviously
no point in arguing. Instead he asked why his opponent's appearance had changed so much during the encounter—why his lance
had grown so unfairly and why his color had changed.

"Ach well, that is after being two questions with different
answers. As to why you saw Sir Fitzgerald's lance grow,[4] haven't I
already explained that? Because he was coming at you so fast,
didn't the tip of his spear come upon you almost as fast as the light
that had left his hand when he was much further from you? You
saw the spear tip almost upon you when it seemed that its bearer

· · · · · · · · ·

4 In special relativity the Fitzgerald contraction is the effect whereby moving objects
   are measured to be shorter. This doe not mean that you would necessarily *see*
   them as shorter however. The Fitzgerald contraction is what remains after you
   make due allowance for the fact that light takes different times to reach you from
   different points.

was himself still some way off. Of course he was actually much closer as, like the arrow that was fired toward you before, he was coming at you almost as fast as the light that told you where he had been. That was why he seemed to come to you so fast. The light that showed him afar off arrived but a short time before he was himself upon you and so his passage from far to near seemed fast indeed. In contrast, the arrow that you saw fired from left to right did not stretch, or shrink, in this way, d'you see. Sure and this is because the light took much the same time to reach you from every point along its path. You might have noted the Lorentz contraction, so you might, but that would be small.

"The change in color is something else that you will have to get used to when you are dealing with fast movement. 'Tis what is called the Doppler effect. Light is a sort of ripple in space, so it is, and blue light ripples more quickly and has a shorter wavelength between successive ripples than has red light. When something moves toward you, it is coming up behind the light that it sends off and so the successive ripples it emits are squeezed into a shorter space: the light has a *blue shift*. By the opposite reckoning, when the light is coming from something moving away from you, then the ripples are spread out over an increasing amount of space and the light is *red shifted*.[5] So you see, in our world things are forever coming at you out of the blue."

As Kevin struggled to sort all of this in his mind, there was a fanfare of trumpets. A crowd of courtiers trooped into sight, led by an elderly chamberlain with a long gray beard. He cried out in a loud, if quavering, voice. "Here comes her Majesty. Bow down to Queen Photania." The Earl sank on one armored knee and bowed his head. A loud noise like a whip crack sounded nearby and another voice cried, "There goes her Majesty." The sharp sound rang out once more and a third voice called, "Here she comes again!" The Earl was swiveling from side to side on his grounded knee, trying to face his sovereign who, unfortunately, was not easy to locate. A vague presence could be seen flitting to and fro, but so quickly it was difficult to see where she was, and she did not stop for an instant in any position.

"Sure and that is Her Majesty, Queen Photania," whispered Puck. "A dainty little thing she is, a mere slip of a thing really. She

. . . .

5 The same Doppler shift is used in a police radar gun. The speed of a car affects the frequency of a radio wave reflected back from it.

is so light on her feet. Well actually she is just so light. Indeed she has hardly any mass, d'you see, and so if she has any amount of energy at all, at all, why she must forever be moving at a speed not far short of the speed of light. She is known as '*La belle dame sans* mc^2,' so she is.[6]

"She is always on the move, and that as fast as may be. She moves so much faster than common sound that the air reacts with shock as it is forced to get out of her way, and she announces her presence with a sonic boom each time she passes. I do not think you would say, though, that she breaks the sound barrier, since for her it is no sort of barrier at all, at all."

As Kevin tried to focus on the flitting figure of the queen, he heard the repeated loud cracks caused by her passage and also a series of high squeaks followed by low lingering groans. The chamberlain approached the young prince and briefly inclined his head formally.

"Prince Kevin, her Majesty says that it pleases her to see you in her domain." The young man thought that he in turn would be quite pleased to be able to see the queen at all, but that did not seem to be an option. He asked why, if the queen had just said all that, he had not heard her himself.

"Oh, it is because of the Doppler shift. You may have noticed how our knights look blue when they ride toward you, whereas they appear red as they depart. The frequency of the light you receive from them is higher as they approach and lowered on their parting. With sound, the effect is much the same, though the effect is exaggerated because sound travels so much more slowly."[7]

There was another burst of squeaks and groans and the chamberlain hastened to interpret. "Her Majesty understands that you are on a quest to discover the nature of motion and that, of course, means that you must seek some understanding of the effect of motion on space and on time. You must encounter the entangled nature of SpaceTime, or 4-space. She invites you to attend at the castle 4-court."

At this, the various courtiers began to move off in a determined fashion, so Kevin followed. The group soon arrived at an area directly in front of the castle. It struck him as a pretty strange-look-

. . . .

6 This is pronounced "em see squared." It is the energy equivalent of mass.

7 You can observe the Doppler shift for sound in the way that the siren of an ambulance or police car will drop to a lower note as it rushes past you.

ing region. Unlike the jousting field they had just left, this court was not clearly set apart, being bounded on all sides by trees and bushes and by various little walls that meandered from the castle right up to its edges. Indeed, they came closer than the edges. As these various features approached, they contracted and in a strange way began to lose their full dimensionality. The prince could see they extended within the court itself, but there they had all merged into a flat, purely two-dimensional surface. Above this surface, and indeed also below it, there were many parallel planes stacked in layers. On each one it could be dimly seen that there were features very similar to those on the principal level.

When the royal entourage arrived another group of servants came staggering out of the castle carrying a heavy throne-like chair and set this up where it had a good view of the court. With much activity they arranged a rich canopy to shade the top of it. As far as Kevin could tell, all this effort was totally wasted, since there seemed no likelihood that the restless royal would ever actually sit in the chair, or indeed pause in any position. As he understood it, this was one of the penalties of having so little rest mass so that, should she possess even a little energy, she was required to travel at near the speed of light, in one direction or another.

Looking more carefully at the scene before him, he was mildly shocked to notice that on the flat planes in the 4-court there were the two-dimensional figures of people as well as of trees and shrubs. On each of the stacked planes their positions were just a little different, and the quaint fancy occurred to him that, if only there were some way of looking at such pictures rapidly in sequence, it would give an illusion of movement. He commented aloud on this curious thought.

"Well now, sure and that is in fact what you are after seeing, indeed it is." His remark had been picked up by Puck, who had appeared at, if not a little below, his elbow. "What you see there," he went on, "is no more nor less than successive slices through time. Things change and move around, so each moment is just that little bit different. You are looking at a view of 4-space, and it has four different directions. As well as the three dimensional space you would expect, you also have the direction of time. You are seeing not only *points* in space but *events*. These are particular happenings that have a definite location in both space and time, so they are. In order for this view to make any sort of sense to your eye, or rather to your mind, the three dimensions of space have been squashed flat to appear to you as but two, though in actuality space is still three dimensional, to be sure."

"Well, it doesn't make much sense to me," rejoined the prince. "I know that I can usually expect to see space, but I do not see time."

"How can you be after saying that?" exclaimed Puck vehemently. "I hope that when you look around you do not think that you are seeing space in some way unaffected by time. You cannot believe that, not after all that I have told you. When you look at something far away, you do not see it at the same time as you see something nearby. When you look into the distance, you look into the past, for the light that you are after seeing has come to you from the past, so it has. It has traveled along the light cone."

As he spoke he gestured toward a glistening cone that rose up from the center of the court. There were in fact two cones. One stretched straight upward into the future, spreading out symmetrically from the center as time advanced. A balancing cone stretched downward, descending into the past and again spreading steadily wider as it became more remote.

The chamberlain had become tired of standing by the throne in a vain hope that the queen would settle there for a moment and shuffled up to them. "That marks the royal road, the path by which those without mass will travel as they fare abroad through the structure of space and time. It is not only light that travels at this 'limiting velocity', of course. Anything that has no rest mass of its own may, indeed must, travel at just this ultimate speed." He paused to fit over his rather prominent nose a pair of small spectacles on the end of a long ribbon. Through these he peered shortsightedly at the cones.

"These two cones in this display join at your position in the center of things. It shows how space and time will appear to you as you stand where the two cones meet," he began. As the chamberlain paused briefly at this point, Kevin took the opportunity to interrupt and point out that he was not in fact at the join of the cones, but was standing here looking at them. The chamberlain sighed as if compelled to enlarge upon the obvious.

. . . .

The light cone

Special relativity may be illustrated by plotting a diagram of four-dimensional space-time. This has three space dimensions plus a direction of time. As we cannot draw or even visualize four dimensions, this means that in practice only two dimensions of space can readily be drawn.

The diagram is centered on the observer and the path that light traces through space and time from this origin is shown by the light cone.

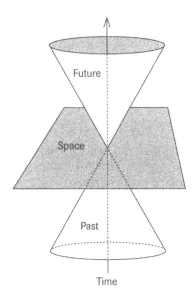

"This display is but a depiction staged for your particular instruction, young man. As far as you are concerned, I am sure that you usually do feel as if you are at the 'center of things.' This represents where and when you are at this moment. It is *your* position and *your* moment, so where the cones meet represents 'here and now' as you see it. You are the star. I shall expound with the assistance of my colleague."

He looked toward the castle. Kevin followed his gaze and saw approaching a young man carrying a bundle under his arm. This new arrival also wore spectacles. In his case the spectacles were small and round and combined with an implausible wispy beard. He had a pocket stitched to the front of his doublet, containing an impressive array of quill pens. He was apparently an example of that comparatively rare type: a medieval computer nerd. Arriving at their side, he held up his burden for their inspection and removed its wrapping. They saw a sculpted head of a dark bronze color.

"Hello, I am a member of the Guild of Automata Artificers," he introduced himself. "This is a genuine bronze head Mark I, as designed by the great Roger Bacon to reveal all the secrets of the Universe. I shall use it to assist with an explanation of the 4-space diagram." He looked expectantly at the head. Nothing happened for a while. Then the bronze mouth opened and a low grating voice said, "Time shall be."

"OK." said the artificer. "That refers to the future light cone. The upper cone that you see there is bounded by all the possible paths that light may take as it moves from you into the future. The region within this cone contains your possible futures, all those events that you might sample without moving faster than light, which of course you cannot do. Along the cone lie all those events where someone or something might look back and see you as you are now. Up the center of the cone is your axis of time." He stopped speaking and again looked expectantly at the bronze head.

There was a pause longer than before, and Kevin thought he detected an expression of metallic obstinacy. Then the head spoke. "Time was" it said curtly.

"This, of course, refers to the past light cone. This is your past, or rather all your possible pasts. Within this cone will lie all the events of your actual past, which must be separated by movements slower than light, as you could not have moved any faster than that. On the cone lie all the events that you might possibly see from your present viewpoint. Anything you can see will be in your past, since it takes time for light to reach you. And now comes the really interesting bit," he said, not entirely convincingly.

The dull features of the bronze head seemed to be in the grip of some intense mechanical emotion. "Time is..." it began gratingly, then without warning burst asunder into a thousand fragments. "Oh sugar!" said the artificer. "It's crashed its operating system!" He despondently began to pick up all the tiny pieces and gather them together in the head's original wrapping.

"I never could be doing with all these gadgets," remarked the chamberlain, who was looking with some disdain at the artificer collecting up the fragments. "Anyway, all the events within the past and future cones are separated from you by *intervals* that are *time-*

like. They are more time than space, and it is conceivable that something might travel through such an interval. The remainder of the 4-space, the region outside the two cones, this is neither past nor future. This is the truly remote: the unknowable. All events in this region are separated from you by *space-like* intervals and no information, no message or messenger, may bridge such an interval."

"Do you mean then," asked the prince, "that there are events in this region that must remain forever unknown?"

"Who said anything about forever? The events I spoke of cannot communicate with you *now*, but if you wait, a signal may eventually reach you. This will be at an event further up your future time line. As you advance into the future, the point where your light cones meet moves with you and the region of your past spreads out to include events that before had a space-like separation. As soon as a light signal may reach you from any event, then it becomes included within your past light cone, and it is possible for you to know about it. You know the saying 'everything comes to he who waits.' Is that clear now?" he finished rather impatiently.

"I see," said the prince, who did see, sort of. "But why do you choose to talk so much of *my* future. Aren't past and future the same for everyone?"

"No, certainly not!" exclaimed the chamberlain, looking at him in exasperation over the top of his spectacles. "Exactly what you will see depends on where you are, when you look, and also on how fast you are moving at the time."

"All right, I can appreciate that the time and place from which I look will affect what I see, but why should it depend on how fast I am moving?"

"Have you not already seen that? I understand you had quite a furious ride in your coming here and noted that time slows down for a moving object. It is also true that for you it has shortened. This is not what you actually *see*, though. Because light takes a laggard path you see a thing some time after the event and the delay will be greater for some positions than for others. You have first of all to allow for that delay, and when you do you find that for people moving at different speeds, space and time really are different. Let me show you what I mean.

"Ma'am!" he said abruptly, addressing the flickering presence of his monarch. "May I request that one of your knights aid us in this demonstration?" A brief squeak followed by a long groan was presumably assent, as he turned to one of the assembled knights. "Sir Lorentz, would you be so good as to gallop off into the distance?"

"Certainly Ma'am," replied the good knight, replying as if to the queen herself. "In what direction?"

"It is no matter. Just go."

This last suggestion was in effect a royal command, even though relayed at second hand, and the faerie knight promptly mounted and rode off. Kevin was struck at how slowly he departed, despite his apparent anxiety to obey the queen. Then he realized that what he saw was, of course, the *light* from the departing knight, and the further away he went, the longer it took the light to arrive.

The vision in the 4-court changed. A new time axis appeared that soared up at an angle to the previous one and was accompanied by a new selection of flat surfaces that denoted space as experienced by the departing knight. These were tilted from the planes that indicated Kevin's space just as the knight's line of time was tilted. Both were folded in toward the light cone.

"There you see time and space as they are for Sir Lorentz. Even when he is still by your side, he inhabits a different time and space merely because of his motion. His time and his space are tilted toward the light cone when compared with yours. The faster he should move the more it would seem to you that his space and time would fold together. Were he to travel at the speed of light, his moments would, for you, last forever, and all his distances shrink to zero. You would observe that for him there was no longer either time or space."

The young prince and the elderly chamberlain stood side by side and looked at the vision in the 4-court, still dominated by the gleaming light cone. The chamberlain spoke again.

"In general, motion is relative. One observer may see something as moving but another may match speeds with it and so see it as stationary. Yet another observer may move still faster yet and then he or she will see that thing as moving in the opposite direction. Such motion is entirely relative. All observers have their own individual time and space. There is one glaring exception to the notion of relativity: if one observer were to see something as moving at the speed of light, then all observers will see it moving at the speed of light.

"You might expect that someone moving toward a source of light and someone moving away from it would see the light as traveling at different speeds. This, after all, is what you see with a running man or an arrow. Because in fact they all see the same speed, this implies that space and time are not as you might expect. They present themselves differently depending on your movement. This illustrates the differences." The chamberlain turned and pointed at the SpaceTime display in the 4-court, now showing two axes of time and two corresponding versions of space

"You observe there how space and time appear both to Sir Lorentz and to yourself. No one of you ever sees *himself* as moving, he just find himself in different places at different times. For Sir Lorentz the axis of time is not the same as for you, since as time passes you find yourselves at different positions. His axis gives all the events that for him will be *here* at successive different times. That is just another way of saying 'where he will be at different times' and shows simply that he moves away from you, and so his axis of time, his position of 'now,' will veer away relative to yours. There is nothing really surprising in that. His different view of space may not be so obvious. Since space and time are so intertwined, his movement away from you will also cause you and he to denote 'now' at different times in any given place. The tilted plane you see before you indicates all those events at different positions in space that he would reckon are happening *now,* and they are not the same places as you would find. The two of you have a different space and time because of your relative motion.

"They are so different that something that is in the future for you may actually be in his past. Consider an event that is far, far away." The chancellor made a gesture at one of his minions who cautiously cast into the 4-court a small golden star that came to rest between the two planes that denoted 'now' for the two observers. "For each of the two of you, your space is tilted in time with respect to space as it appears to the other, simply because of the relative movement. The distant event indicated here is below his space plane, his plane of 'now', and so for him it is in the past. You can see here that the same event is above your plane of 'now' and so is in your future."

· ·

Space and time for different observers

Everyone sees light as moving at the same speed. This requires that observers moving relative to one another see space and time differently. The orientation of the space planes may be made clearer by looking at the axes shown lying in these planes. They are along the direction of Sir Lorentz' motion for the respective observers. Note that, for the moving

observer, Sir Lorentz, both this space axis and the time axis appear to
the 'stationary' observer, Kevin, as having folded toward his light cone.

A distant event may lie between the two space planes and so be in
the future for one observer and in the past for another.

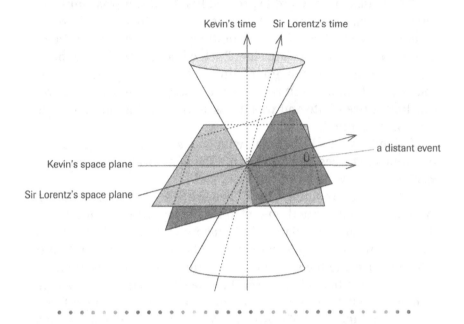

"Surely that cannot be!" exclaimed the young man. "Whatever
distortions you say there may be in space and time, I can be sure of
the order in which things happen. When I experience one event, I
will know perfectly well whether or not I can remember the other.
I find it hard to believe that different people will see past as future
or future as past."

"They do not." The chamberlain again looked severely over his
spectacles. "In fact, the point is that they do *not* see them. Not until
much later. It is only in the isolated space-like region outside the
light cone that such ambiguity exists. That is not to say that such
events must be forever unknown. Just wait and you may see them
when their light reaches you, but by then they are all some time in
your past. It is when different observers see the light that they *infer*
the different times."

Once again the frequent booms of the queen's passage were
accompanied by a shrill squeak sinking to a lingering groan as her
words limped through the air.

"Her Majesty feels that you have heard enough of the nature of time and motion, and it is now time for you to move back toward your home. Puck will escort you as far as the boundaries of her domain."

The prince bowed and expressed his thanks. It was simple to bow to the chamberlain, but he was hard put to know where he should properly direct his bow to the queen, so he bowed in all directions to be on the safe side. Puck came up on his little pony and told him to climb up behind. The young prince did as requested, though he felt rather self-conscious as he towered over the diminutive sprite and was thankful that his legs did not quite trail on the ground.

"Right now, 'tis back home with you, so it is." The ill-matched twosome (or threesome if you count the pony) rode away from the castle and along the forest track.

"Well now, my young friend," remarked Puck conversationally, "you have some quare thoughts to take away with you. You know that all movement is relative. You may move relative to something else, so you may, but you cannot say whether it is you or it that is moving. Sure and the very question is meaningless. This makes it all the stranger that all, however they may be moving, will see light as moving at the selfsame speed. None may move faster than this.

'Tis a strange sort of truth, this, but as it is found to be so then surely the relation of space and time must be distorted in unexpected ways from how you might expect. You have a slightly different time and space from anyone else who is moving at a different speed. You cannot say whether you are moving and they at rest or contrariwise, but you may certainly say whether the two of you are moving differently. You should note, however, that the distortions of SpaceTime that occur are less than they may appear to your eye. Your eye, poor thing, can only see light after all—and light that reaches you from a distance only tells you of past history. When you see moving objects as they were in the past, why naturally they are in different positions. When you look out into the far distance you look far into the past, so you do, and the farthest distance you may see is in the depths of space. It is between the stars and galaxies that you may find the distances and the speeds such that these distortions of your perception become really important."

They rode along in silence for a short time as Kevin mulled over distinction between what was and what he saw. Then Puck broke the silence to speak of another aspect of motion.

"You have learned that energy and momentum are important aspects of motion, and they, too, differ from what the common folk hold them to be. Though velocity may not rise past the allowed limit, both momentum and energy may grow to any value. As speed approaches its greatest possible value, it is as if the very mass does rise.

"The other thing you now know," continued his little green guide, "is that energy still has a considerable value even when there is no motion. Anything with mass has 'rest mass energy.' Indeed you are after finding that mass and energy are the very same thing.

"And that is about it. That is all you know and all you need to know—or close enough. Now I must leave you," he said abruptly as they approached the edge of the wood.

"Can't you come just a little farther?" asked Kevin, who had become rather attached to the assertive elf.

"That he may not!" broke in another voice. "By Her Majesty's express command he is forbidden to accompany you any farther." Near the edge of the path, and looking quite out of place, there was a small desk at which sat a severe-looking fairy in a formal dark suit. On the desk there was a notice that read "The Puck stops here."

Regretfully, Kevin took his leave of Puck and continued alone along the path to his home. He soon found the small valley, entered the castle, and made his way into the king's presence. There he told his father everything that he had discovered about

motion. "Well done, my boy. You have completed your quest; you are now my rightful heir. Although," said the king as he turned to look through the castle window at the dense forest that bounded the kingdom, "I now rather wish that I had sent you to learn something about tree felling instead...."

INTRODUCTION TO
the second tale

"I have told you a tale of momentum and energy, of space and time and of motion in general," remarked the Storyteller to his audience as he settled more comfortably on his seat. "Now you should be in a position to listen sensibly when I talk to you of stars and of their dramatic movements through the vastness of space. Or rather," he added carefully, "you will be in such a position when you have heard something about the nature of the matter from which stars are made.

"How may you hope to understand the behavior of a star throughout its life if you do not have any understanding of the matter that has been concentrated to form it? The way that stars behave depends on the way that matter behaves, and this in turn reflects the nature of the particles that it contains. How stars develop, live, and die, and the way they manufacture the elements of which we are all made. These all depend on the behavior of atoms and nuclei in the appropriate conditions. It is this behavior that has created the myriad stars we see.

"Before we attempt go further, I shall tell you a tale that tells of atoms, of electrons and nuclei, and the strange quantum

behavior that dominates them. I shall tell you the tale of 'Snow White and the Particularly Little People.'

"Snow White was a princess who lived with the queen, her stepmother. The queen was jealous of Snow White and one day arranged that the royal hunt master should take her into the woods, supposedly on a picnic but in reality to kill her. The hunt master could not bring himself to do this and told Snow White that she should flee, as far and as fast as she could...."

the second tale:

SNOW WHITE

and the

particularly

little PEOPLE

(Into the deep basement of our world)

Snow White fled through the wood, running, running ever deeper into its dark interior. As she ran, seeking to lose herself in the forest depths, jagged tree branches tore at her dress. Suddenly she lost her footing and tumbled into a steep-sided depression in the ground. She rose shakily to her feet and looked around, realizing that she had succeeded in her aim. She *was* lost.

As she wondered what to do about this, she noted four signposts nearby. This was encouraging, as a signpost is generally a welcome sight when you are lost. She went over to look at them.

On closer examination they did not seem as helpful as they might be.

One pointed toward the sky and read:

GRAVITY: LOOK UP.

Another had arrows pointing to every side and read:

ELECTROMAGNETISM: LOOK AROUND.

A third pointed toward the ground and read:

STRONG NUCLEAR INTERACTION: LOOK DOWN.

The fourth and final signpost seemed to be totally inward looking and this read:

WEAK INTERACTION: LOOK OUT!

All in all they did not strike the princess as being very helpful in her present, or indeed in any, circumstances. She was looking at them with some distaste when a small voice said, "Those refer to the basic interactions of the world, you know."

The girl looked quickly around, but could not see anyone. "Who are you?" she cried out.

"I am one of the Little People. Look down by your feet." Snow White looked as directed and saw, peering over the toe of her dainty slipper, a tiny little man.

"What do you mean?" she asked.

"Why, those notices refer to the four basic interactions that bind all the matter in the Universe. Without them there would be nothing that you could recognize as matter or as Universe. Of course there would be no 'you' to do the recognizing, either.

"On the largest scale you see the effects of gravity, binding together the stars into great whirling galaxies. On a slightly smaller scale gravity moves planets, such as the very Earth on which we stand, in their orbits around their parent Sun. Look up!"

Snow White looked up and saw, through a gap in the leafy cover of the trees, a view of bright stars in the sky above.

"Here on the surface of the Earth, however, you live in an electrical world. Gravity has but a small effect. It might not have seemed so when you tumbled into this deep hollow, but your tumble did not last long, for you soon landed on the ground and the ground was easily able to support you. Here on Earth you are much more concerned with the material objects of the world around you, the liquids and the solids that make up your surroundings and even make up your own body. You may have to take care not to fall, but if you do fall the consequence is normally only a bruise. You do not disintegrate into your component atoms.

"It is the electromagnetic interaction that binds solids and gives them such strength that the attraction between a few nearby atoms can easily resist the force exerted by the gravity of the whole Earth. Come and look around with me and you will see the effects of these atomic interactions."

• •

Electromagnetism is our sort of interaction

We are creatures of electromagnetism. Electromagnetism is the interaction that binds atoms together as well as providing the internal interactions that hold the electrons inside atoms. It is responsible for chemistry and it provides the light we see by and, one way or another, the nerve impulses that carry the vision to our brains.

• •

Snow White was astonished to see that upon these words, the little man expanded abruptly until he was as large as she. This dramatic transformation made her feel uneasy, as if she also was in some way *changed*, and she looked around for reassurance. The trees arching overhead looked much the same, though somehow more remote. The distant stars also looked remote, but no more than they had before. Her immediate surroundings were different, though. Where before she had been standing in a grassy glade, she was now surrounded by thick vegetation at shoulder height, great flat green leaves that sprang from the ground on all sides.

"That's right my dear," remarked the Little Person as he followed the trend of her thoughts. "You have shrunk down to my scale and are in among the blades of grass. As you move around on this level, you will find that gravity is of less importance than before. Come along." He led off at a smart pace and the girl followed. What before had been small stones strewn about were now huge boulders and even small rocky prominences, but to her surprise she found that she could climb them rather well and that when she leapt down the far side she suffered no damaging impact, even from several times her own height. She paused for breath by the side of a small puddle in the dirt.

"You see that it is as I told you. When you are small the effect of gravity is proportionately less as you climb or fall. It becomes less important the smaller you get."

· ·

Gravity is for the big boys

Gravity is important for planets and galaxies, but the smaller an object is, the less it is affected by gravity. An ant can lift many times its own weight. This is not particularly because it is an ant but because it is *small*.

· ·

"Indeed it does!" interrupted another small voice. This time Snow White looked down immediately and once again saw a tiny man peering over the edge of her dainty slipper. "Hello, I am one of the Even Littler People," he said without waiting to be asked. "Come down and join me and you will experience even more intimately the triumph of the electromagnetic interaction over gravity." Before she could think to resist, even had she known how to, the girl found she had shrunk down to the size of this new acquaintance. The Little Person to whom she had been speaking now towered above her into the sky. He gave a friendly wave of farewell and strode off with the earth vibrating from his receding steps.

"Right!" said her new companion. "Let's be on our way. We will go over the lake, I think. Ignore the notice."

The little princess looked around her and saw that the splash of moisture on the ground had become a wide lake stretching away. On the bank there was a notice that read 'Please do not walk on the water.' Ignoring this, the little man took Snow White's hand and led her out onto the surface of the lake. She was so confused by the recent events that she did not resist, and it almost seemed reasonable when she found that her feet did not sink into the water, but rested in little dimples that appeared on its surface. The water behaved as if it were a thin jelly covered by a rubber sheet. Although her feet did not sink, it was very difficult to walk because the dimples changed shape as she moved and tended to pitch her forward on her face. She was rather afraid of what might happen if she fell full length upon this liquid surface.

"Surface tension," remarked her companion.

"I beg your pardon. What do you mean?" she responded.

"It is the surface tension of the water that is supporting you. All the molecules within the water are pulling on one another, and this makes it seem as if the water has an elastic skin. The molecules do not pull anything like as hard as they do in a solid, I admit, which is why a solid will support you even at your normal size, but it is not until you shrink this far that water will bear your weight. The notion of walking on water should not be completely novel to you. I dare say you will have seen insects running over the surface of ponds. In fact here comes one now."

Snow White looked around and saw a great armored shape come scurrying over the surface toward them. The insect, for such it was, came to a stop alongside. "Nice day!" it remarked in a friendly tone[1]. "I hope you do not mind my remarking that you are not making a very good job of walking on the water. Comes of having only two legs I suppose; bad idea, that. Would you like a

- - - - - - - - -

1 A talking insect is of course an example of the "pathetic fallacy." If famous poets can make use of this, I do not see why I should not.

ride to the shore?" The massive insect was looking at her
with what was probably a
friendly expression, though it
was rather hard to tell with
someone who had a mouth
like a fairly well-equipped
machine shop.

Snow White accepted
readily. She was finding it *very*
hard to keep her balance. "Thank you very much" she said and
struggled onto his broad plated back with a bit of assistance from
her new guide.

Their transport set off across the surface of the pond. Six wide-
spread articulated legs gave a smooth and stable ride, each one
creating a little dimpled cup in the water as it descended. About
halfway across the pond the tranquillity of their ride was disturbed
as a brief flurry of rain fell from the sky. Snow White had never
been all that fond of rain, but she had never experienced it like
this. Huge drops like hogshead barrels enclosed in plastic sheets
fell on the surface of the water. As each hit the surface, the water
splashed up in a ring of gleaming spikes, like a great liquid crown.
The little princess was terrified.

"What happens if one of those hits us?" she asked.

"Oh we shall probably sink. Don't worry though, it doesn't
happen very often," replied her mount. "Look, the shower is over."
Sure enough the drops were no longer falling, and they made their
way without further incident to the far shore. That is perhaps a
rather extraordinary word for the opposite edge of a tiny puddle,
but it seemed like a shore from the girl's viewpoint.

On dry land, Snow White and her companion jumped off the
back of the obliging insect with a word of thanks and continued
inland. The gigantic grass stalks that surrounded them looked
strange to her, each surface covered with a sort of lumpy cell
structure. She was in fact seeing the cells of the living tissue. Most
strikingly, the grass no longer looked green as it had before. There
was very little sign of color anywhere; everything had a strangely
indefinite look. When Snow White remarked on this, she was told
it was because she was now so small compared to the wavelength
of light that she could not see in the normal way.

"In order to see fully how different things are on the small
scale, you have another step to take."

"Right, and now is the time sure enough." The voice had come once again from the region of her toes, but in very little time she had shrunk to stand face to face with the speaker. He looked much like her other companions, dressed in tights and a simple tunic. "I take it you are another Little Person," she said.

"Oh no, the Little People are much, much larger. I am one of the Exceedingly Tiny People. I am here to show you a scale of being where you may see how the classical regularity that the large-scale world appears to possess is built upon a basement world of quantum uncertainty."

"So, can you tell me, if I am too small for my eyes to work properly, how *can* I see?" she asked plaintively.

"Well, strictly speaking you can't. Surely you cannot believe that you are still actually *seeing* anything. At this size light has a wavelength that would seem to you more like high frequency radio. You are not seeing things anymore than you are breathing the surrounding air. The molecules of oxygen would now be far too large for your lungs. You are, as you ought to realize, an allegorical figure acting as part of a purely illustrative exposition, so please do not ask awkward questions.

Snow White looked around (or experienced, or whatever) and found there was very little left in her surroundings that she could recognize. There was little indeed of which she could make any sense at all. The world about her was now seen to consist of an enormous number of separate objects, vague fuzzy spheres. She couldn't tell what they were, as all the objects were now hazy and uncertain. It seemed to her as if they were all in continuous movement, but it was hard to tell. Choosing a direction at random, she started to walk when suddenly she spied something that looked quite normal, except that in comparison to the rest of her surroundings, normality was particularly unreasonable.

Across her path was a barrier, a long striped pole hinged at one end and carrying a disk at its center that bore the sign

$\hbar$

This was Planck's Constant, the symbol that conveyed the measure and character of the quantum realm. A man wearing a very neat gray uniform approached and raised one hand commandingly. "Stop!" he demanded firmly.

"Why should I stop? What is this?" asked Snow White rather crossly.

"This is a customs and immigration post. You are about to enter the realm of the quantum, and so you must have a customs check to see that you understand our customs and comply with local regulations," responded her tiny companion. "Don't worry. Just answer the Customs Inspector's questions and you should be on your way in no time.

Snow White obediently turned to face the prim official. "I have to check that you are not smuggling in any inappropriate attitudes," he informed her. "To start with, do you by chance have a particularly large amount of common sense about your person?"

"Well, I should hope so. I do not think that you can complain about common sense!" responded the princess hotly.

"Oh, it will have been fine where you come from," replied the official soothingly. "Very good and useful in fact. It is a sort of distilled residue of all the experiences that have been common in your life, but it is not really legal tender here. The trouble is that you will not have had much, if any, experience of reality as it is found at this quantum level, and so your common sense will lead you astray. It is not a reliable guide here."

"Well, if it isn't, what is? If common sense and previous experience do not provide a good guide, what can guide actions here?" exclaimed the girl in exasperation.

"Listen to the Customs Inspector," suggested her companion. "It is part of his duties to tell you the customs of the locality, how things are done here."

"Since you ask me," remarked the Inspector on cue, "in my capacity as a customs inspector I shall try to tell you what is customary here. Here things are guided by a *wave function;* indeed in a sense the things *are* their wave functions. A wave function extends over space and has the form of an amplitude with a continually varying phase. The amplitude gives the probability of each and every different outcome that is possible: the likelihood of each path down which a system may go."

"I am sorry," said Snow White, in the tone that suggests it is the other person who should be sorry. " I do not understand what you are saying. What is a wave function and what is a phase and how do they guide anything?"

"A wave function or an amplitude was in the past sometimes called a *pilot wave* because it was seen as somehow guiding a particle."

"How does that work?" asked Snow White. "Anyway, even if it works for particles, how about everything else?"

"Everything else *is* still particles," retorted the Customs Inspector, holding up a manual titled *Regulations for Permissible Composition of the Universe*. "All matter, everything you can see in your normal life and even the light by which you see it, is composed of particles. As the saying goes: 'Particles, particles, all is particles.' Particles are described as fully as they may be by this wave function or amplitude. The full amplitude tells you pretty well everything about the particle that you can hope to know. As far as you are concerned, you might as well say it *is* the particle. A wave function is in some ways a bit like sound or a wave on water. For a water or sound wave the amplitude is a displacement from an initial state of rest. For the wave function it is a displacement from a state of *non-being*. The amplitude gives a measure of the degree to which the particle is actually *there* at any given position. The intensity of the wave function at any point, as given by the square of the amplitude (the amplitude multiplied by itself), gives the probability that the particle might be detected at that point. In a way it measures the degree to which the particle is *present* at that point. You get particles present to a greater or lesser degree at various points, you know."

. . . .

Particles and wave functions

On the atomic level, particles are not simply tiny hard billiard balls. Their exact position becomes unclear and the best description we can give is in terms of a *wave function* or *amplitude*. This is a rather diffuse wave-like thing from which you can get the *probability* that the particles might be detected in a given position. That is all you can say, really. Physical theories may speak only of what you might hope to observe.

. . . .

"That is ridiculous!" retorted the girl. You cannot have the same thing present at two different places at the same time, let alone 'various points'." The Customs Inspector stared at her with a particularly blank official expression.

"Are you sure that you are not carrying more than the permitted allowance of common sense?" he asked. "You cannot of your own knowledge say what an electron should or should not be doing as you will have had no experience of single electrons. I expect that you have never even been introduced to one. Anyhow, the electron wave function will have different values at different places, but there will be a whole range of places at which it is not zero, and so the electron is, in some sense, partly present at all those places. There is some probability that it might be *found* at one of those places. The way in which the amplitude varies is determined largely by the phase," he continued dispassionately.

"And what is this phase?" demanded the princess determinedly.

"That is a bit hard to state clearly without mathematics. You wouldn't happen to have brought much mathematics in your luggage along with all that common sense would you?" asked the Inspector hopefully. "No" the girl replied curtly.

"Well, in a sense phase is a matter of attitude, of cooperation. In classical theories of physics you just had probabilities for the different things that particles could do. Probabilities do not have phases, they all add the same way. If there were more ways that a particle could end up in a given condition, there would be some probability corresponding to each of the ways. The probabilities would all add together, so that as a result it would be *more* likely the particle would end up in that condition."

"That sounds fair," said Snow White. "Yes, it's only common sense, really."

"Perhaps so," replied the Inspector. "It is in a way a bit like a contest against a team of soldiers." As he spoke, two teams came running onto a nearby area, wearing contrasting colors of shorts. You could tell that the members of one team were soldiers, as they also wore uniform tops with cross belts and brutal steel helmets. The two teams faced one another, and then the soldiers' team was belatedly joined by a host of virtually identical reinforcements. Their opposition quailed as this horde bore down upon them and easily forced their way through.

"The soldiers are like the probabilities that I spoke of in that they are uniformly oriented in their purpose, and the more there are, the greater the overall effect. When you speak of amplitudes, however, then you must realize that their orientation varies. This is more like a contest against a team of lawyers."

Snow White saw that the team of soldiers had vanished and a team of lawyers came running out. She could tell that they were lawyers as, above their colored shorts, they wore short gowns and legal wigs. The members of the original team now faced these new opponents, and as before a horde of reinforcements came running out to join the team of lawyers. Snow White observed that the members of this augmented team all began furiously to argue with one another and their opposition was able to run past them virtually unopposed.

"This illustrates a property of phase, in that some of the lawyers opposed their fellows while others took intermediate posi-

tions. This is the case with quantum phase where some amplitudes oppose others, and these tend to cancel out. It is even more complicated than this, in that the phase of a particle is forever cycling around through complete cycles of orientation. It is as if the lawyers were forever changing their minds and arguing on the opposite side, and also were all changing their minds at different rates, so that sometimes they agreed with a colleague and sometimes were fiercely opposed."

"That sounds totally chaotic," remarked the girl.

"Well, it can certainly get complicated. The addition of amplitudes is known as interference. Sometimes they are in agreement and they add. This is constructive interference. Sometimes they are in opposition and cancel one another out, which is naturally called destructive interference. Mostly their relation is somewhere in between. Interference like this is responsible for most of the structure you see in your world.

. . . .

Probability, phase, and interference

Interference is the most striking quantum effect observed for particles on an atomic scale. Even in a classical theory of the world you might have to admit that, at the atomic level, you had no chance of *measuring* exactly what was going on. You would be forced to speak of probabilities for different conditions, but at least you would be sure that if there *extra* ways in which something could happen, then it was *more* likely. Even this turns out not to be true. Wave functions have some sort of phase or orientation to one another. If they are *in phase* then they will combine, if they are *out of phase* they will cancel.

In practice you see such cancellation in interference effects. Where you get light or other particles reaching the same part of a screen by different routes, you observe light and dark bands from constructive and destructive interference. This is an ever-present effect and cannot be explained without phase.

. . . .

"I suppose I should try to demonstrate how it works." He produced a bag and thrust his hand inside, then withdrew it to cast something into a sort of enclosure that was conveniently nearby. The girl saw this was now filled with a hazy cloud, pronouncedly more intense in some places than others. She saw that, although the cloud was motionless, the intensity varied greatly from place to place, sometimes rising to a peak of intensity, sometimes dying away to almost nothing. The assembled lawyers had all stopped arguing and were watching this new display with great interest.

"I have injected an electron, so you are seeing an electron wave function in the cavity. Even though there is only one electron present, its wave function has many components that correspond to the options open to it. The waves are reflected back from the walls and all have phases. For a single isolated wave it does not matter very much, but you do not often get isolated waves. When two or more waves interact, this orientation controls how the over-all amplitude will behave. Sometimes the individual waves support one another and the overall amplitude is greater, sometimes they are completely out of sorts and disagree so completely that the result is nothing. These are the extremes of *constructive* and *destructive* interference."

"How do you have so many waves when you said there was only one electron?" asked the little princess. "Surely if you have one electron, you only have one of these waves or whatever they are."

"Not so. I must repeat that a wave function represents or describes what a particle might be doing. Usually it might be doing many different things and there will be a wave function present for each possibility. There must be, as the rule here is *whatever is not forbidden is compulsory!*" He waved a neatly gloved hand toward a notice board on which this slogan was prominently displayed. "You may perhaps know the lines:

All the world's a state,
And all the various particles merely functions.
They have their phases and their amplitudes;
And each one at a time has many waves

"There will be a wave present for every reflection of the electron from a wall and these waves can all interfere. The pattern of peaks and valleys that results is called a *standing wave* as the pattern does not move, being fixed by the boundaries of the enclosure.

"Now a wave such as this has some sort of amplitude everywhere, and I have told you that in a sense the wave is the electron. You might ask how an electron can actually move if it is already everywhere." Snow White supposed she probably would have asked that if she had thought of it.

"Motion is well-known to possess both momentum and energy, and these are found to be intimately tied in with the wavelength and frequency of the wave function. There may be no reason evident why they should be, but you may accept without question that, that is what is found to be so.

• •

Wave functions and kinematics
The frequency and wavelength of wave functions are related to the energy and momentum of their corresponding particles.

Frequency, ν, is proportional to energy, E:

$$\nu = \frac{E}{2\pi\hbar}$$

Wavelength, λ, is inversely proportional to momentum, p:

$$\lambda = \frac{2\pi\hbar}{p}$$

(The factor $2\pi\hbar$ is a constant value. The symbol $\hbar$ you saw earlier on the Customs Inspector's barrier.)

• •

The higher the energy of the particle, the higher the frequency of the wave; the higher the momentum the shorter is the wavelength. These quantities are invariably so related," he said firmly, opening his manual to glance at an appendix. "It may perhaps surprise you," he continued in his calm and supremely unsurprised way as if he were still reading from his manual, "that a wave with a definite energy and momentum will spread uniformly over all space. There will be no distinguishable feature in any position and so nothing definite that may appear to move. One region may appear different from another only when waves combine and the resultant interference may give a peak of intensity in one place and little elsewhere. Let me demonstrate."

He reached again into his bag and produced what the girl could only presume to be another electron. He did not add this to the one in the enclosure, but cast it upon the ground. The electron spread out over the ground to give a thin vague mist with a locally intense region at one point. This moved off enthusiastically and bounced off various obstacles, much as if it were a conventional particle.

"There, you see that there is a definite peak in the probability that moves from one position to another. You now have visible motion of that peak–the position where the electron *most likely* is. If it were detected, then it is almost certain that it would be found somewhere in that region. The narrower the peak of probability, the more precisely you can say where the particle actually is–that is to say, where it would be discovered were you to look."

"That must be nonsense!" stated the princess firmly. "A thing must surely be wherever it is and nowhere else. If you detect it somewhere, then that can only be where it was."

"Are you certain?" responded the Inspector calmly. "You should not be. In this realm the only thing that is certain is a certain uncertainty and the positions of particles follow the Uncertainty Principle. Perhaps you might say that you knew where something was, but you still cannot say exactly where it is now and certainly not where it will be. The main purpose of physical understanding is to let you say how things will turn out; to know what *has* happened requires only memory. If you wish to say where a particle will be in the future, you can only talk of its probability distribution. If the distribution has the form of a narrow peak, then the particle will be most likely somewhere in the region of that peak. That is all you can say, really.

"This localized region of high probability is called a wave packet. Even though you have only one electron, its state is described by the sum of many waves, as I have told you. A narrow peak like this requires that all the many waves should *interfere constructively* at the peak but should get rapidly out of step with one another to give massive *destructive interference* close by. If the waves are to get out of step in so short a space, there must be a large spread in their wavelengths and consequently in their momenta. This is the Uncertainty Principle.[2] A small uncertainty in position for a particle means that there is a large uncertainty in its momentum, and vice versa.

"The same sort of relation applies for time and energy–the shorter the time the bigger the uncertainty in the amount of energy present. Take this entry visa," he said abruptly, handing the princess an official looking slip of paper. "This will permit you to stay here for an uncertain amount of time. Have a good interaction with our environment," he added, saluting her and stepping back

· · · · · · · · ·

2 The Uncertainty Principle is discussed in more detail in my books *The Wizard of Quarks* and *Scrooge's Cryptic Carol.*

from the barrier, which swung upward to allow Snow White and her companion to proceed on their way.

They traveled only a short distance when they came to a small thatched building. The walls were heavily carved and ornamented and outside the decorated door hung a sign that read "Atomic Artificers (Elements assembled by hand to your requirements)." There was obviously nothing to do but to go in. Snow White stooped to pass through the quaint doorway and found herself inside a long room. If the outside of the building had seemed heavily decorated, the inside looked as if it had been the home to generations of mad whittlers. Everything that could conceivable have been carved was deeply incised and the walls bore an unreasonably diverse array of cuckoo clocks.

Down the center of the room ran a long table at which sat seven bearded little men who peered near-sightedly at the tasks in front of them. Most of them were concentrating intently, though one at the end appeared to have gone to sleep and another was sneezing repeatedly.

"What are you doing?" Snow White asked the one nearest to her. He was seemingly too shy to reply, but his neighbor answered happily.

"We are assembling atoms," the Artificer said. "See here." He dipped his hand into a little bag, much like the one used by the Customs Inspector and took out electron after electron, fitting each carefully into the object in front of him. This was rather fuzzy in appearance, as the princess had by now come to expect, but there were nonetheless signs of a quite intricate pattern. For the first couple of electrons it looked a simple ball shape, but as each extra electron was added, it grew more complicated. Each new electron added the shadow of a new complex splendor within the overall structure.

"You will notice," remarked her companion, "that each electron that is added must go into a different state. Sometimes this makes a big difference in the energy of the electron, sometimes no difference at all, but in every case the states differ in some way."

"Why is that?" asked Snow White. "Is it because all the electrons differ from one another?"

"Not at all! Quite the reverse. You see the fact is that electrons *want* to be individualistic. They want to be different from their fellows, but this is difficult for them because they are in fact all exactly, identically the same. That is not to say that they are very similar or even almost identical. They are *exactly* the same: *totally*

identical. There is no way for us to tell them apart. There is no way
for Nature to tell them apart, either. What happens is that each
electron must be in a different state from all the others. No two
electrons may be in the same quantum state, an effect known as
the Pauli Exclusion Principle."

"And that is how they come to be distinguished from one
another?" suggested the princess.

"Well no, unfortunately not. You see, as electrons are identical,
they cannot be distinguished from one another by any means.
There is no way of saying which electron is in what state. Any one
of them might be in any state, and it is a founding rule for the
quantum condition that there must be an amplitude present for
any condition that is allowed–'What is not forbidden is compulsory' as the regulation goes. Consequently there is present an
amplitude for each electron to be in each state, though there is not
actually more than one electron in any state.[3] You see effects of this
all over. The reason you do not fall though the floor in your normal
condition is not because of any interaction that pushes you away. If
anything, the electrical interactions might tend to pull your feet
into the floor. You do not sink because the electrons in your feet
cannot be in the same place as the electrons in the floor, and that
forces you out.

"That all sounds like complete nonsense!" exclaimed Snow
White indignantly.

"Yes, fun, isn't it," said the Artificer happily. "It means that
every electron I add has to go into a different state from the ones
before, and this often means that it must have a higher energy, as
you can see."

"Of course she can't see that!" growled his neighbor grumpily.
"You know that ordinary people do not see energy. Try not to be so
daft."

"Oh, of course. I am sorry my dear. Here, have a pair of these
E-focals. They will allow you to see how things are distributed in
energy, rather than in space. I am sure that you will enjoy the new
experience."

He took a pair of strange looking spectacles from the table and
handed them to her, beaming broadly. She put them on and saw

· · · · · · · · ·

3 Some justification for the Pauli Exclusion Principle is given in my books *Alice in Quantumland* and *The Wizard of Quarks*.

that there was an area set into the lens that was somehow different from the rest. She looked at the atom through this part of the spectacles and, though she could still see the little men normally through the rest of the glass, the atom itself now appeared to her as a set of levels, set one above the other. These, she assumed with a burst of inspiration, must represent states of different energies. Sure enough, the electrons already present were in the lowest levels and as another electron was added, it fitted into the next empty level above.

• •

Energy levels in atoms

When particles are confined, as are the electrons in an atom, they are restricted to a set of distinct possible states. In general, each state gives a specific energy to an electron in it, and these are the energy levels in the atom. Excited electrons may shift from one level to another, and when they do, their energy changes by the difference in the levels. Energy conservation leads to the emission or absorption of photons of the appropriate energy. Atoms will consequently emit photons of specific, distinctive energies, resulting in a unique line spectra for each type of atom.

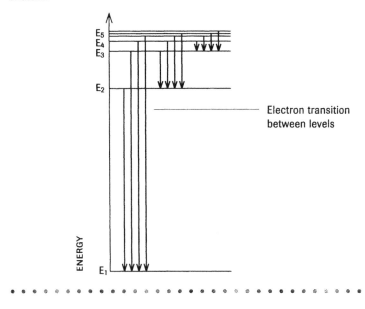

Electron transition between levels

• •

As the Artificer continued to add electrons, Snow White noticed that for each one added there was a little bright flash from the vicinity of the atom. Then an electron stuck briefly in one of the higher levels that were empty. It stayed there for a moment and then tumbled down with another bright flash as it fell. When she commented on this to the friendly Artificer, she was told that what she saw were the photons that were emitted when the electrons fell to states of lower energy. Because it is a general rule that energy cannot be destroyed, when an electron loses some energy, the energy must go elsewhere and usually it is carried away by a photon. Photons are what we see as light.

"There are a lot of photons around inside atoms anyway," said her informant cheerily. "The electrons are held in place by electric attraction, and that is due to virtual photons, as you must know," he finished with an inappropriately broad grin.

"Ach, man, yer an idiot. The little lassie will know nothing of the sort!" said his grouchy neighbor. "Look here girlie, I doubt ye know what virtual particles are. They are those whose existence is permitted solely by the quantum fluctuations in energy. For a short period nature allows an uncertainty in energy, and the shorter the time, the greater the amount of energy that may be borrowed. Photons, which are the quantum particles that make up what you know as light, may be given off by any particle with electric charge. Electric charge is one thing that electrons have aplenty. If an electron has more energy than it needs or is good for it, then it may emit a photon to carry off the rest. Even if it does not have

energy to spare, it may emit a virtual photon by taking advantage of the short-term uncertainty in the total energy. It may borrow its energy from the general fuzziness of the world. It cannot keep this energy for very long, but in its short life the photon might travel from an electron to some other charge and such *exchanged photons* give rise to electrical interactions."

"Just so," agreed her previous informant happily. "As I said, when an electron drops to a lower energy level it emits a photon with an energy equal to the difference in the energies of the two levels. The energies of the levels are different for each type of atom. The light an atom gives off will consequently contain photons from a distinctive set of possible energies. The light given off by atoms of a given type is spread out over different colors or frequencies, usually called the spectrum of the light, and it appears as a number of sharp lines. Look for yourself"

Snow White looked for a while at the light coming from the atoms as electrons changed from one level to another. She was still wearing the E-focals and so she saw the light as a series of narrow sharp lines, some much more intense than others, some a bit fuzzier, but all distinct and forming a complex sequence.

"These lines correspond to the energy differences of the various levels in the atoms and give a unique 'fingerprint'. You may identify the type of an atom, whether it is oxygen or carbon or iron or whatever, solely from the light given off."

"All right now everyone," broke in another of the Artificers, who seemed to fancy himself in charge. He clapped his hands together as he tried to organize the others. "Come along then, let us start to assemble the atoms we have produced so far."

All of the Artificers began to lay out on the table their collections of atoms and to put the atoms together. As the atoms were brought close together, they joined to their neighbors.

"Atoms often stick together. It is known as bonding. Sometimes one atom has a space for another electron to fill completely its outermost energy level with electrons. Another atom may have a single electron all on its own in a higher level. When the second atom gives its electrons to the first, the overall energy of electron binding is reduced, and Nature likes to free energy in this way.[4] The result is that one atom now has a negative electron charge too

· · · · · · · ·

4 The release of binding energy to give kinetic energy of some sort is favored by thermodynamics. Very much favored. This is discussed in my book *Scrooge's Cryptic Carol*.

many and the other one too few, so the simple electrical attraction of opposite charges will hold them together. In other cases the atoms will simply *share* electrons, and this also glues them together. "[5]

As Snow White watched, more and more atoms were brought out and fastened together. Some combined to create great crystal arrays of metal. Some combined to give huge organic molecules that fitted together into complex masses of material. More and yet more were produced, the mass of material soon overflowed the table they were working on and surged toward the decorated wall. As their whole workshop was merely an allegorical construct, it provided no hindrance to the rapidly expanding physical structure. This large physical object, composed of many atoms, passed without hesitation beyond the room but could still be sensed expanding at ever increasing pace into the real world beyond. More and more and more–Snow White saw inconceivable numbers of atoms added to form the growing object that stretched away into what was, from her present viewpoint, an unimaginable size. At last, as she peered out into the cosmic dimensions beyond her current atomic scale, the girl could see, in the far astronomic perspective of a remote distance, the vast but unmistakable outlines of a macroscopic physical object–one that had the unmistakable appearance of a cuckoo clock.

"And is all this indeed built on the way that electrons fit into the energy levels within atoms?" she asked wonderingly.

"Oh yes, it all comes from the way in which electrons that are contained in a small volume must fit their waves precisely within the space available and so are forced into a limited selection of wave patterns that correspond to different energies."

"What does contain the electrons?" asked the girl. "I did not see any sort of container outside the atoms you were constructing."

"You wouldn't expect to see one because the container of an atom is in its center. It is also very small. The electrons are all held within the atom by the electrical interaction from a tiny but massive positively charged nucleus that pulls upon the negatively charged electrons and keeps them in position. Atoms are governed by electrical interactions, as you will have been told, and these pull the electrons down toward the nucleus, much as stones and princesses are pulled down toward the Earth."

• • • • • • • •

5 See carbon bonding as described in *The Wizard of Quarks*.

"If that is so, why don't the electrons fall down until they hit the nucleus?

"That is because the electrons are too light to fall so far. If an electron were to fall right into the nucleus you would know where it was. Agreed?" The girl could not but say yes. "In that case the *uncertainty* in its position would be very small and so the uncertainty in its momentum would be large. You cannot have a large uncertainty in momentum without having a large momentum, and consequently a large kinetic energy. For an electron to get close to the nucleus, it would need more energy than it could get from the electrical attraction and so it can fall only so far.[6] If you calculate the separation of electron and nucleus at which the electrical attraction is balanced by the tendency to spread out associated with the Uncertainty Principle, you will find the result is about the size that atoms are found to be."

The princess passed up this invitation as she did not care for mathematical calculations and instead asked, "OK, so the electron may not be able to get too close to the nucleus, but what is the nucleus?"

"Well it is, as I say, at the center of an atom, such as this specimen here. It is very small…"

"Don't you bother listening to him," came a new and rather small voice. "What would a great oaf like that know about anything on the scale of the atomic nucleus? Just you come with me."

Snow White looked down, by now quite expecting to see a small figure peering over the toe of her dainty shoe. She didn't. She couldn't really see anything. She was, as far as she could tell, being addressed by a tiny speck of dust on the floor. As she was wondering how there could ever be any meaningful communication across this size gap, the speck grew rapidly and rather dizzyingly to be revealed as yet another diminutive man, with a particularly large, friendly smile. She was not overly surprised when she looked around for her previous companion and could see nothing but a vague hazy shape that seemed to fill the whole sky–and that was only his foot.

. . . .

6 This argument may sound a little vague, even dubious. However, if you work through a calculation of the wave function for an electron in an atom, it comes out to have broadly the size of an atom. This is not so surprising as the Uncertainty Principle is a consequence of the wave-like properties of particles.

"Hello," her new guide introduced himself. " I am one of the Truly Infinitesimal People. You may call me Truly if you wish. You were asking about atomic nuclei. Well, there's where you can find out about them."

He indicated a featureless, rather barn-like, building that was situated nearby. Its appearance seemed totally implausible deep within an atom, or indeed anywhere else. It had a large overhead sign that read:

ATOMS "R" US

Below this was a slightly smaller sign stating: THE SECRET IS IN THE NUCLEUS!

The miniscule companions made their way to the store, for such it amazingly appeared to be, and went inside. They were greeted by ranks of display cabinets and bright sales posters on all sides. A banner strung across the entrance read, "Whatever you want, the answer is elementary" and "Elements, elements, all is elements!"

Scattered around the various showcases were bright notices:

"USE CARBON FOR YOUR ORGANIC PRODUCTS."

"TRY A MIXTURE OF OXYGEN AND NITROGEN—IT WILL BE A BREATH OF FRESH AIR."

"TREAT YOURSELF TO SOMETHING DIFFERENT—TRY A RARE EARTH. ON SPECIAL THIS WEEK IS YTTRIUM."

As they were looking at the display a figure rushed forward to greet them.

"Good day Sir and Madam; so pleased to meet you. I am a Sub-Atomic Sales Person, at your service. I can assure you that you have come to the right place for all your atomic needs. I see that you have been looking at our special offers on ready-to-use nuclei. We carry the whole range you know. Any atom that you might want, you have but to take the appropriate nucleus and add electrons.

"Alternatively," he remarked, with a slight change of tone, "as a true connoisseur you may like to handcraft a nucleus of your own. We carry self-assembly kits for every type of nucleus you could desire, and our ingredients are as stable as any you will find."

Truly took it upon himself to interrupt this flow of welcome. "I should like you to show this young lady how one may assemble an

atom. In particular, would you please tell her about nuclear construction and the components you use?"

"Of course, of course. As you have already seen, you make an atom by simply adding electrons to a nucleus. Indeed, all you have to do is put the nucleus in a cloud of electrons. We store the electrons over there." He pointed to the far end of the cavernous building, which contained crates so enormous that that just a few of them were sufficient to fill most of the available space.

"Why are the boxes so large?" asked the princess. "Are the electrons very large then?"

"Well, no. Actually, they are very small in themselves, but they are very poorly localized. Electrons are very light you see. They have a very low mass compared with most other particles. If you tie down their position, that is to say, you make their position uncertainty very small, then the momentum given to them will be large. You may know that momentum is given by mv, mass times velocity, so if the mass is small then velocity must be large. An electron confined within a nucleus would have very high speed and escape from any containment, much as a fast space ship will escape from the Earth. You can see that it is difficult to put electrons into small boxes" he concluded.

"Anyhow," he continued. "You can see that because the nucleus has a positive electric charge, it will attract electrons because they have negative charges. This electrical attraction binds them to the nucleus. The fact that the electrons are then crammed into a fairly small space will affect the electron wave functions and result in all the atomic levels and such-like that make atoms behave as they do. You have probably heard this described already."

"Oh yes," said Snow White. "I think I know something about that. It sounds the same as I saw the Atomic Artificers doing. They were adding electrons one after another to make an atom."

"Well, there you are. Though, as I said, you do not have to add the electrons deliberately. Just put the nucleus in with the electrons and its electrical attraction will pull them in. This goes on until there are enough electrons in your new atom to balance the positive charge on the nucleus, then no more electrons are attracted and you have a neutral atom. That is all there is to it, really, though our products do carry a warning that building or splitting atoms is not really suitable for very young children.

"It is as our sign says, 'The Secret is in the Nucleus.' The type of atom you get depends on the nucleus that you start with. You will want our premium nucleus construction kit," he stated confidently. He reached into a cupboard and took out a decorative box. He opened it to reveal a couple of large packets and a tiny pot. One of the packets was colored a cheerful red, the other a dull green.

"Here you have the component parts. They are basically protons and neutrons, and the red package holds the protons while the neutrons are in the green one. From the point of view of making the sort of atom you want, the protons are the most important. They have positive electric charge, the same amount but the opposite sign to the charge on electrons. The number of protons in a nucleus determines how many electrons the nucleus can capture. This fixes the sort of atom you get, whether it be oxygen or carbon or iron or even uranium. It all depends on the number of protons in the nucleus."

"Why do I need the neutrons then?" asked the girl.

"Well, to begin with it is a bit like what people say about climbing mountains—because they are there. Neutrons do exist and are fairly stable. Protons and neutrons are both affected by the strong nuclear interaction, the interaction that is able to bind particles within a tiny nucleus. Consequently, neutrons will be present as well as protons, like it or not. The Pauli Exclusion Principle allows the protons and neutrons to fit independently into the lowest energy levels, since protons and neutrons are not identical to one another. Protons are certainly identical to other protons and neutrons to neutrons, but a proton and a neutron are different. Protons and neutrons can fit quite independently into the lowest energy levels available.

"The presence of neutrons helps to make nuclei more stable. As I said, the protons have the advantage of having positive electric charge, so they can attract the negatively charged electrons. However, they also have the *disadvantage* of having positive electric charge, which means that each one of them will repel all the other protons. The positive and negative electric charges serve to bind the electrons within an atom, but the only effect of the positive charges carried by the protons in the nucleus is to make the whole thing fly apart. You need something else that operates between protons and neutrons, the *nucleons* as we refer to them collectively–something that serves to bind them all together. This interaction mustn't attach to the electrons, of course, or it would distort the energy levels of the atoms. What you need is an interaction stronger than the electrical interaction, one that acts between nucleons but does not affect electrons.

"What you need is the 'Strong Nuclear Interaction'." The Sales Person paused impressively and held up the small pot. "This comes along with the kit. It gives you some bosons that the nucleons can exchange to glue the nucleus together, in the same sort of way as exchanging photons provides the electrical attraction. These only exchange over short distances, so you might say that they stick the nucleons together only where they touch. This feature makes it quite unlike the electrical interaction, which extends over long distances–to the limits of the atom and beyond. Inside the nucleus the strong interaction can win out against the repulsive electrical forces between the positive charges on the protons, but it is a close thing. The neutrons serve to separate the protons a bit and so reduce the electrical effect. A nucleus with just two protons does not exist, it would be torn apart by the repulsion of the charges. A nucleus with a proton and a neutron does exist; it is deuterium, a heavy hydrogen atom. In fact, even a nucleus containing two protons and a neutron can and does exist.[7] Neutrons

• • • • • • • •

7 It is called Helium3.

are quite effective at offsetting the repulsive force between the protons' charges."

"And how do you go about getting these particles to join together so that they will form a nucleus?" asked the princess. "Do you just let them drift together, as you say you can with electrons when you have a nucleus already present?"

"No, that wouldn't work in this case. The electrons attach to the nucleus because the two have opposite electrical charge. The protons all have the same electric charge and so they push one another away. They just wouldn't come sufficiently close together for the short-range strong interaction to work. You need really fast-moving protons that can make it past the repulsion to the point where the strong interaction can grip them. You want a nuclear furnace. One with a temperature of a few tens of millions of degrees centigrade will do. Even that isn't hot enough for the thermal velocities of the protons flying about to allow them to make their way through the electrical repulsion, far enough for the strong nuclear interaction to take over. They may get close enough however for there to be some chance of their *tunneling* through the gap that they do not have sufficient energy to bridge. It's a quantum thing," he confided.

"And where would you get such a furnace?"

"Oh, the core of any sufficiently large star ought to do." Before Snow White could make any comment about the availability of such a location in the domestic environment of the normal consumer, the Sales Person continued, "The electrical barrier isn't the only problem for the creation of nuclei. There is usually also a shortage of neutrons, and you cannot make stable nuclei without them."

"I thought they were provided in roughly equal numbers."

"True, but I am afraid you have to read the small print." He held up the green package to show a label that read 'CAUTION, contents should be used within fifteen minutes'. "Neutrons are unstable, you see. They decay steadily and after a few hours almost all of them would be gone. It is because they are susceptible to the Weak Interaction, you know."

"I take it that is different from the Strong Nuclear Interaction."

"Oh yes, quite different. It is another thing entirely. It is much weaker to start with, and it has its own unique effects. It can change particles and cause them to decay. A particle may decay if there is something lighter for it to decay into. Energy will be conserved overall and so any reduction in mass will release the equivalent amount of energy as kinetic energy. Nature likes kinetic energy.[8]

"A neutron will decay to produce a proton, an electron, and a very light uncharged particle called an antineutrino. Protons and neutrons are both heavy particles, a couple of thousand times as heavy as an electron, but the neutron is slightly the heavier of the two. Its mass exceeds the mass of all three of the final particles added together. In this case the decay of one particle to give three is still able to release a little energy, and so it can happen."

"If neutrons are unstable like that, I do not see how they can be of any use in building nuclei and atoms. Surely atoms need to last for a long time?"

"Ah, that is the cunning of it. Neutrons are warranted to last for billions of years provided they are used as recommended and built straightaway into nuclei. They are stable enough then. It is because of the binding energy, you see."

"So what is binding energy?" persisted the princess.

"It is the energy that a particle loses when it is bound to others, when it falls into a potential well. A well in your experience is a hole in the ground and if something should fall down it, it would fall faster and faster, until it hit the bottom. There it loses its kinetic energy in some way, mostly by making a 'splash' sound and warming the water slightly. It would need to have kept all that kinetic energy if it were to soar out of the hole again, but it has lost it and so it is bound. Particles are in much the same quandary when they are held by an interaction. They would need energy to break loose, but they do not have it. The energy they lack is called their binding energy.

" Now energy is just mass, so if a neutron has lost energy, it has lost mass. When it is bound inside a nucleus, it is no longer heavier than the three particles to which it might have decayed. When it is bound within an appropriate nucleus, the neutron is quite stable," he finished soothingly.

"Well, I do not see why you should use unstable components at all. It seems a bit substandard to me."

"Oh, but it is not," returned the Sales Person with a winning smile. "We use nothing but the very best constituents available. In fact

• • • • • • • • •

8   There are a lot of ways to distribute kinetic energy among different states, and the more possible states there are the more likely is the process that results in them. This is discussed in Chapter 2 of my book *Scrooge's Cryptic Carol* and is the heart of thermodynamics.

we use the *only* constituents available. Come and talk it over with our Quality Control Technician."

They made their way across the great barn-like building, decorated with forceful exhortations to buy this or that product, and passed through an inconspicuous door into a small room with a workbench along one side. Seated on a chair in front of this was a man who was wearing, balanced on his nose, a sort of metal framework similar to those optometrists sometimes use for patients to try out lenses. On this were perched a considerable variety of unfamiliar devices.

"This young lady would like to know about the constituents we use in our nucleus kits," stated the Sales Person, thereby unconsciously demoting Snow White from Princess to Lady. "She has some doubts about their stability."

"OK," returned the technician, peering up at them past an array of filters and other instruments. "Well, the obvious constituents are the proton and the neutron, of course." He took one of each out of little drawers and set them on the bench in front of him. They stared at them for a few minutes. The neutron thereupon decayed to give a proton and an electron, which fell off the table. There was, of course, no sign of the almost undetectable antineutrino produced.

"I take it that is what is worrying you. The nucleons are not the only strongly interacting particles, but the others you might think of are not a lot of help. For example there are the antiparticles."

"What is an antiparticle?"

"I am just about to show you." He picked up a tool on the end of a heavy armored cable and held it over the bench. He pressed a button on the tool, and there was a brilliant flash and two particles appeared simultaneously. One was clearly a proton, the other looked similar, but in some way subtly different.

"That is an example of *pair production*. With a suitable energy input you can create the rest mass, for new particles. Usually the particles must be created in pairs, as they have other properties that must be balanced. The proton has a positive electric charge and so its companion anti-proton has a negative charge." He pressed the button again and again two particles appeared. One was a neutron and this time its companion was even more subtly different. "The neutron does not have any overall electric charge and so the anti-neutron does not differ in that way. There are however differences.[9] The two particles are not the same and they do complement one another."

He pushed the neutron and anti-neutron together and they disappeared in a great flash of light, with photons of very high energy rushing away in opposite directions. He then took a pair of neutrons out of one of his drawers and pushed them gently together. They just sat there side by side, not doing anything in particular apart from the usual quantum spread in fuzziness.

"Anti-particles all have the property that they may annihilate their complementary particles. As this is allowed it will usually happen, since they produce lighter particles and so release kinetic energy. You can see that anti-protons and anti-neutrons, if mixed with the normal particles, would not be a good basis for stable nuclei. There are other particles you can make." He produced a succession of flashes with his energy source, each time creating new particles unfamiliar to the girl. "There are plenty of different strongly interacting particles, so why do we use the nucleons? Well, all of these new types are even less stable than the neutron, so they are not a lot of use for building stable nuclei. The proton and neutron are the lightest two in their family, so there is nothing lighter available to which they may decay. I admit that the neutron is just heavy enough to decay into a proton, an electron, and an antineutrino, but this is marginal and usually cannot happen when the neutron is bound within a nucleus. The neutron is stable enough then, and its presence is very valuable in keeping the protons apart so that their electrical repulsion won't tear nuclei apart.

"All of the other strongly interacting particles are probably present to a degree, in that the uncertainty relation between time and energy will allow enough energy to be borrowed to create particle/anti-particle pairs, but only for a very short time. Such virtual particles and particle pairs would be present as they are present even in a vacuum.[10] To some degree it is a moot point whether

· · · · · · · · ·

9 They are different although they have the same mass and electric charge—neutrons are made from quarks, and quarks are in some way inherently different from antiquarks. The main evidence for a difference between particle and antiparticle comes in noting that they can only be created in complementary pairs. Some more detail is given in *The Wizard of Quarks*.

10 See *Alice in Quantumland* or *The Wizard of Quarks* for discussion of the vacuum.

these particles are present or not, since they are all made of the same ingredients. They are composite objects and they are all made from quarks.

"Here, look at the electrical structure of protons and neutrons." He placed on Snow White's nose a framework similar to the one he was wearing and positioned a couple of disks in front to her eyes. "These are not actually lenses, of course. They are conceptualizers. They help to show you, not what you can see, but what is actually there–a very different thing. This set shows you electric charge, how particles appear to the electrical interaction. I have a set for each of the basic interaction types, though the ones for gravity are not very helpful to us and are anyway far too heavy."

Through the spectacles that were not spectacles the girl saw that the proton was colored in a vivid shade of positive electric charge. It was brightest in the center and faded away on either side. The neutron, even though it was neutral, was also visible to her electrical perception. Its center also was glowing brightly with positive charge, but this was surrounded by a wide halo of negative charge so that, overall, the neutron was indeed neutral. It was clear, though, that it had some sort of structure.

The Technician took a pair of very small tweezers from another drawer and began delicately pulling the nucleons apart. From them, and from the various particles scattered about, he picked out a number of tiny objects and spread them on a tray. Snow White noted that these were still not actually separate, but were bound together by colored strings.[11] He set them out in three columns side by side, with two quarks below one another in each column.

"There you have the quarks: the six basic varieties. I have arranged them according to electric charge and family. There are three families and in each there are two particles of different electric charge. The three different families are known as *generations.*" The top element in each column glowed with positive charge, the lower more faintly with a negative hue. This was the case for each of the three columns, which looked much the same. "That is the set of quarks as seen from the viewpoint of the electromagnetic interaction, for which electric charge is all-important. If you look at them as viewed by the strong interaction then you see that all the quarks are quite distinctive. They have different flavors."[12] As

· · · · · · · · ·

11 There is more about quarks and color in my book *The Wizard of Quarks.*
12 I'm afraid that flavor really is what they call the property that distinguishes the quark types.

he said this, the Technician replaced the filters with another set. The particles in the different rows and columns all looked completely different from one another. "You are seeing how the quarks appear to the strong interaction, an interaction that can tell all the quarks apart. Things look a bit different though when viewed through the weak interaction. Now the flavors are somewhat blended" He replaced the filters with another set. The distinction and clarity that had belonged to the various quark types, from the viewpoint of the strong interaction, had faded. The top and bottom positions in each column merged slightly and the quarks in the top row combined members of the various flavor families.

"As far as the weak interaction is concerned the different flavors of quarks are not distinct, the weak interaction does not clearly tell them apart. The weak interaction selects out something other than flavor and the types of quark it sees are a mixture of flavors. The weak interaction can blend flavors together within the same state and cause a decay from one flavor of quark to another. The weak interaction may cause the quarks in one generation, as the different columns are called, to decay to a lower and lighter form.

"There are also six leptons, he remarked taking out a tray that contained six particle grouped in much the same way as the set of quarks, in three columns, or generations, of two particles. "Leptons are particles like the electron. They may be arranged in three generations, like the quarks. Each generation has a negatively charged particle and its own neutrino. The charged versions are respectively the electron, the muon, and the tau. The latter two are heavier and decay quite readily, so only the electron, from the first generation of leptons, plays a significant role in atoms."

"Why do you have the other two generations then?"

"I have no idea. They just come with the set." He put away the tray of leptons and turned his attention back to the quarks. "It is essentially the same for the quarks. Only the quarks in the first generation, the 'up' and 'down' quarks as they are called, are important in nuclei. The others, the 'charm' and the 'strange', and the 'top' and the 'bottom', they are not. Not really."

"Could you make atoms with them, those charm ones for example?"

"I suppose so, in principle. Normally these other quarks do not last long enough to get built into atoms."

The Technician took some quarks out of a drawer that was neatly labeled 'charm.' He combined them together with other quarks in a way that the girl could not quite follow. He added electrons to form atoms and then took another handful of quarks. The resulting mass of atoms grew apace, much like the great clock that she had seen the Atomic Artificers construct earlier. This time, however, as the object grew, Snow White found herself to be growing alongside it, rising upward from her miniscule scale. She saw the atoms forming into larger structures. Nearby was the regular array of a simple metal, further off the atoms were forming the complex, convoluted structure of organic material. As she grew further she realized that the metal was a silver buckle, the organic material a fine tunic. Swiftly, or so it seemed to her, she recovered her normal size, once again standing in the grassy hollow from which she had begun her journey into the atomic underworld.

She was not alone. Beside her stood a young man—well dressed, handsome and captivating. "Hello," he said in a thrillingly seductive voice, " I am Prince Charming."

INTRODUCTION TO

the third tale

"*Of course, Prince Charming would not really have been made up from 'charm' quarks," mused the Storyteller. "They are unstable and decay with such a short lifetime that he would have exploded in a tiny fraction of a second. Had he actually existed, he would have been formed from 'up' and 'down' quarks like everyone else. Of course as he was entirely fictional, it is pointless to debate the physical composition he did not possess.*

"*Anyway," he continued abruptly, "you have now heard both about motion and about matter, so you may usefully begin to contemplate the Universe at large. The Universe, as you know, is full of stars, and the nature and distribution of these is a wonder to contemplate. All the stars, all the great galaxies that contain hundreds of billions of stars, all the super clusters of galaxies, the great web of matter that fills the Universe..."*

"*You haven't told us about any of these things!" protested Rachel.*

The Storyteller stopped short in mid-flow. "No, I haven't, have I?" he admitted. "I had better remedy that. I shall tell you a tale that depicts the huge extent and variety of the Universe around you. Quieten down then, and I shall tell you the tale of 'Ali Gori and the Cave of Night.' Our previous stories have been of a prince and a princess, but now I shall speak of a rather ordinary young man."

As his audience settled down reasonably attentively the Story-teller took a deep breath and began. "Ali Gori was a young man of no remarkable birth but some ingenuity. He lived in a far-off desert land, and one day was making his way through that desert...

the third tale:

ALI GORI

and the

CAVE *of* NIGHT

(*consider* the *heavens*)

Earlier that day Ali's landlord had reminded him that it was necessary from time to time to pay his rent. Now he was making his way
carefully across the barren desert, on the lookout for circumstances that might make him some sort of offer he could not resist.
He spied in the distance a small group of camels moving along
roughly parallel to his path. Gradually their courses converged
and he could see that their riders were quite well dressed. Probably merchants he thought, or maybe scholars. There had been
quite a spate of wise men in these parts recently. He preferred
merchants, though, as they had more interesting baggage.

As they were going in much the same direction, he cautiously
shadowed them as they rode into a deep valley and drew up before
a cliff face. They all dismounted and marched up to the cliff, which
was in deep shadow. Their leader raised his arms in a dramatic
gesture and declaimed something that Ali was too far away to hear.
After a short pause there was a deep grinding noise and the group
disappeared, apparently straight into the cliff. There followed a
long pause, during which Ali crept closer to the scene. Eventually
the robed figures appeared again, and this time he was close
enough to hear what was said. It did not help him much. One
turned to another and remarked, "I still say it was a type-one

supernova," after which they all
mounted their camels and rode
away.

When they were out of sight,
the young man crept down to
stand where they had stood, below
a rough and weathered cliff face.
Above the level of the young man's
head there was a massive, jutting
proboscis of stone, and two heavy
slabs of granite surmounted it,
with deep pits beneath. Underneath it there was a wide crack in
the cliff. To Ali's mind this had the
appearance of a mouth to comple-

ment the craggy nose and eyebrows suggested by the other features. He was smiling at this fancy when the crack opened wider
and a voice like rocks rubbing together addressed him.

"Well, don't just stand there. What is the magic word?" it
demanded in tones of distant thunder.

"Please?" suggested Ali hopefully. The flinty eyebrows drew
together in a deep frown and Ali prepared to run as rapidly as he
could. Then the rocky visage cleared and the mouth spoke again.

"No one has ever said that to me before. It has always been
'you must answer to the magic word' or 'I am your master, hear
me and obey.' I think I rather like 'please.' OK, you can come in."
The stone face crumbled inward in a cloud of dust and rubble, like
a pile of rocks collapsing into an underground cavern. When the
dust cleared away there *was* a cavern leading deep underground.

"Well, nothing ventured, nothing gained," Ali muttered tritely
to himself. After his successful encounter with the petrified and
petrifying guardian of the portal, he was not prepared to pass up
this opportunity. He marched down the rocky passage with a sort
of cautious swagger, and soon the narrow tunnel opened out into a
great circular cavern. He stared open-mouthed at a cave of wonders, a cave of a billion, billion stars. The floor area was huge, but
what really caught his attention were the walls, hung with
Heaven's embroidered cloths. There were the blue and the dim
and the dark cloths, enwrought with golden and silver stars.
Mostly there were the dark cloths, for this was the Cave of Night.
Overhead the cave was roofed with a vast dome, supported on
twelve thick pillars that lined the walls and bracketed the star-

flecked hangings. The wide concave surface of the ceiling was black, with star upon star crowding its surface. Bright stars and dim stars, some relatively close but many unthinkably distant, fading into a remote peppering of tiny, scarcely visible lights.

Ali stared in awe at the myriad stars that blazed overhead. In places, he thought he detected pattern in their display and seemed to see significant groupings. As he lowered his eyes, he noticed a series of niches around the walls, alternating with the pillars. In the niches were what he took to be statues, and they formed a motley collection. There was Sagittarius and Aquarius, with bow and water pitcher. More daunting was the massive armored shape of Cancer with huge claws held in front, and of Scorpio with tail menacingly erect. Pisces lay in a heap quite close to Libra, the scales. Under the heading of Virgo a young woman watched them both with a cold gaze, perhaps considering the purchase of a meal. One alcove was labeled Leo, but was empty as the Lion was not there.[1] Ali circled the hall and stopped by a recess labeled Gemini. Within this he could see twin figures, rather fat ones at that.

As he was gazing at the pudgy shapes he was startled by a voice coming from one of them. "If you think we're waxworks," he

• • • • • • • •

1   He was absent escorting Dorothy on her way to visit the Wizard of Quarks, in my book of the same name.

said, "you ought to pay, you know. Waxworks weren't made to be looked at for nothing. No-how!" The speaker had "Castor" embroidered on his collar.

"Contrariwise," added the other, who was marked "Pollux," "if you think we're alive you ought to speak."

"I am very sorry," answered Ali politely. "Tell, me," he continued in as casual a manner as possible, "this looks an interesting sort of cavern. Is there any ancient treasure or that sort of thing lying around?" he asked hopefully.

"You've begun wrong!" cried Castor. "The first thing to do on a visit is to say 'How d'ye do?' and shake hands. The two brothers gave each other a hug and held out the two hands that were free. Ali did not like shaking hands with either of them first, for fear of hurting the other's feelings, so he took hold of both hands at once. The next moment they were all dancing around and together went spinning out of the cavern and down a side passage.

The two brothers were fat and soon out of breath. "That is far enough," Castor gasped out, and they left off dancing as suddenly as they had begun. "I don't know about treasure," panted Pollux, "not nohow, but there are various antiquities in the museum there." Ali saw that they had stopped outside a door in the rocky wall, and that it bore the legend "Museum of Antique Cosmologies." Cautiously he opened the door and sidled in, followed by the heavenly twins.

They entered into another large cavern, this one dark and shadowed and crowded with dim shapes draped in dustsheets. "What is all that?" asked Ali hopefully.

"Why, those are old cosmologies. I should think this is the largest collection of abandoned cosmologies you will find anywhere," answered Castor smugly.

"They don't come much older. Nohow!" added Pollux, moving over to a great mildewed chest and raising the lid with much groaning of rusty hinges.

The two stout little figures leaned over the edge of the box on tiptoe and lifted out... Ali was not initially at all sure what it was. There was a flat sheet balanced on four little elephants that in turn stood on the back of a large turtle.[2] On the sheet were coastlines and rivers and tiny mountains. A miniature Earth, in fact. Over the

• • • • • • • • •

2 This was one early notion of how a flat Earth might be supported. I refer you to Terry Pratchett's *Diskworld* series of books.

Earth was a great shell, like a round transparent plastic dish cover. This was painted on the inside and pierced with many little holes so that light could shine through.

As Ali watched, a bright ball of fire rose from one side of the disk and soared round beneath the dome, eventually sinking into the Earth at the end of its course. As it moved, the dome of heaven rotated also, carrying with it the fixed stars as a mysterious pattern upon its surface. From the place where it landed, the tiny Sun apparently tunneled back under the Earth, because later on it rose again from the same spot as before to begin its passage across the heavens. Underneath the dome Ali could see a number of other bright balls that were moving around, travelers across the face of heaven.

"That's a pretty old one," remarked Castor. "It don't work too well though. Folk started to measure the Earth's surface and they found it wasn't flat. Nohow!

They had to build themselves a new cosmology with a round Earth and one that made some allowance for the planets–the wanderers across the heavens."

As if on cue his brother had in his turn wandered off and was tugging at a dustsheet draped over a large object. The dustsheet hung down and touched the floor on all sides. With an extra tug he managed to pull it off and revealed a round globe, marked with the Earth's continents and streaked with clouds. Around this circled the Moon, Sun, and several planets, all of these orbiting around the fixed Earth. Closer inspection showed that the circling celestial

balls were mounted on thin crystal spheres that carried them around. The movement of the spheres was accompanied by a distant sound of harmony, the whole effect being rather like a heavenly clockwork music box.

"That one's pretty old, too," remarked Pollux. "Greek workmanship I believe. If you look closely at the whole edifice though, you can see how people have had to mend bits of it from time to time. Just look at those epicycles!"

Ali looked wisely at the construction for a few moments, than gave in and asked what an epicycle was.

"They're all over it," Castor moved forward to enlighten Ali, pushing his brother to one side with an elbow in his rather well-padded ribs. "There's one on Mars–look, the little red planet here. You can see a little plastic wheel fastened to the celestial sphere and the planet is carried round on the rim of the wheel.

"They had to cobble this cosmology up a bit for the usual reason. It wasn't the same as what people actually saw. Nohow! When they watched some of the planets move round the sky in front of the fixed stars, they saw that usually they would move a bit faster than the stars, but occasionally they would reverse and go slower. Moving round on the rim of the epicycle fixes that, so it does. Sometimes the planet moves along faster than the rotation of the sphere that carried the axis of the wheel; half a turn later and it moves against the sphere's rotation. The whole thing kept being made more complicated as people watched the heavens more exactly."

Throughout this speech Pollux had been holding his side and grimacing. Now he went over to another shrouded shape and tugged the cover off. At first sight it looked much like the previous cosmology, but it only took a moment to realize that the central place in this version was held by the brilliant Sun. Around this circled the planets, with third one out the blue ball of the Earth encircled by its own satellite, the Moon.

"At first this heliocentric version, with the Sun in the middle, didn't do much better than the older one. The Sun is now fixed in the middle and the planets, including the Earth, circle around it. The Moon is permitted as an exception, as it travels around the Earth, allowing the Earth some remnant of special privilege. One difficulty with this model is since the Earth moves around in a circle, the stars should look slightly different when the Earth is in different positions throughout the year. The makers of this model could only offer the feeble excuse that perhaps the stars were so far away that any such *stellar parallax* was too small to be seen. Nohow!

"This model did take front place, though, when Galileo started to look through his telescope," Castor remarked as he trotted off to a large armoire set against the cavern wall and took out the long tube of a brass telescope. It was much longer than he was and he struggled to carry it back to Ali. Pollux bent over and Castor rested the heavy tube across his twin's shoulder.

"Look, look!" the brothers cried. "Look at the phases of Venus, the second planet." Ali obligingly applied his eye to the lens and observed the little planet as the two fat little men shuffled round to keep the telescope on target. As it circled around the central Sun, he saw how the light on it changed. When it was on the far side, almost the whole visible disk was brightly lit. When it was well to one side or the other in its orbit, only half of its visible face was lit, the farther side being in shadow. When the planet was on his side of the Sun, then he saw nothing but its dark shadow side.

"This regular cycle of light side and dark side that you see when you look into the sky can only mean that Venus is moving around the central light. This heliocentric *solar system* didn't need epicycles either. Nohow! See what the Earth is doing there." Ali could see that the sphere carrying the Earth around the Sun was moving faster than the one Mars was on. As he watched, the Earth came racing round and passed Mars on the inside.

"Think what you would see if you were on the Earth there," commanded Pollux. "As you passed inside the Martian orbit you would see that planet appear to drift back against the background of stars. You have the effect observed without needing epicycles."

He reached out and rapped a chubby knuckle against the outermost crystal sphere. "These still gave a problem though. They got in the way of anything that went through the solar system, and some things did. Look!" he added, gesturing to the side. Trailing its long glowing tail, a comet approached the compact system and blithely sailed through each crystal sphere in succession as it made a single turn around the Sun. There was a faint tinkle as it crossed each one, and the space began to fill with little broken fragments. "Comets were thought to be omens of disaster. They are certainly a disaster for models that need to have supports to hold the planets. Sir Isaac Newton saved the day, or rather the night. All done in accordance with his Principia."[3]

• • • • • • • • •

3 In his book *Principia* Newton set down his thoughts on Mechanics, celestial or otherwise.

Both of the little stout brothers spun round to face a door masked in shadows. Each dropped abruptly to one knee and held his arms out in a dramatic gesture. Through the doorway came a prim-faced man in a long wig, dressed in knee breeches and a frock coat. He marched up to the waiting group and addressed them confidently.

"What holds the planets in their orbits you may ask. Gravity, that is the answer. Universal Gravitation. The planets are held in their courses by the gravitational attraction of the massive Sun, in just the same way as an apple falls from a tree under the attraction of the Earth's gravity." The Philosopher rather self-consciously took an apple from his pocket and tied a thread to it, then began to whirl it round his head on the end of the string. The apple flew outward and the string tightened until it was almost horizontal. "You see that the tension in the string is opposing the apple's tendency to fly outward and away, as it would if the string broke. In the same way planets are held in their orbits by the attractive gravity of the Sun."

He walked past them toward the end of the underground storeroom, to a wide opening at one end. The others followed until they stood side by side at a balustrade that extended across it. Beyond this was a great empty space, dark as night and bounded by a remote stellar sphere. Before them in the middle of this dark space floated a great blazing orb: the Sun. Their new companion took from his pocket a little ball of some material and hurled it toward the Sun. It missed and flew by at one side, sweeping round in a tight curve and then continuing to circle around the central light. Other planets followed in succession. Ali was able to observe that the third had the familiar cloud-striped blue of the Earth and the fourth was the small red globe of Mars. The next would not fit into a pocket and came from a deep satchel that the philosopher carried. This planet was so massive it took considerable effort to propel into position. The next one was not much smaller and had a rather attractive system of rings encircling it. As each planet settled into position, Ali could see that many came with their own attendant satellites orbiting around them. The most obvious was the Earth's own Moon, but most planets had their attendants, held by *their* gravity.

"The farther they are from the Sun, the slower the planets will orbit. Because the force of gravity weakens with distance from the Sun, then at a great distance it can only hold a planet that moves rather slowly. At the same time the circumference of an orbit will increase the farther out are the planets, so the outer planets take much longer to complete their orbits. Their years are longer than ours."

"What happens if a planet isn't moving at the correct speed for its distance from the Sun? That must surely happen sometimes."

"It happens often. It always happens to some degree. If a planet is moving too slowly for its distance from the Sun then it will fall inward under the attraction of gravity. As it falls it speeds up, but this means it now has some motion toward the Sun. It is not moving in the correct direction to orbit neatly in a circle and continues to fall under gravity until it is moving too quickly for a circular orbit at its distance. Not only gravity is involved, but also the *conservation of angular momentum*. This depends on the component of its velocity, in a direction at right angles to a line from the center of the Sun.

"There comes a time when the value of this *component* needed to give the appropriate angular momentum at that distance from the Sun is as great as the *total* speed the planet has gained in its fall. This means the planet is now in fact travelling at right angles to a line from the Sun. This is its closest distance of approach and the planet subsequently swings out from the Sun, losing speed until it arrives back at its starting distance. From this position it once again falls back. Overall the planet will move in an elliptical orbit with the Sun at one focus."

The representation of Newton paused and looked rather satisfied with himself. "All of the crystal shells have now gone from this picture. There is no need for any celestial scaffolding and the solar system is largely empty, with no obstruction in the path of comets passing through. As for the sphere of the stars, that is shattered completely." He took something else from his pocket and hurled it at the surrounding celestial sphere. As it struck, the sphere shattered and the stars that dotted its surface were hurled outward so that they came to rest spread throughout the immensity of space.

"There are the stars, no longer a familiar wallpaper that drapes our local nursery, but each one a sun in its own right, spread throughout space and time. Absolute space, in its own nature, without relation to anything external, remains always similar and immovable. Absolute, true and mathematical time, of itself, and from its own nature, flows equably without relation to anything external.[4] In this infinite space and time the stars are spread uniformly and forever." The philosopher finished with the note of absolute conviction belonging to someone who knows that, once you have thought things through and however long this may take you, you are bound to end up agreeing with him.

The twins began exchanging mischievous glances and nudging one another. "Come and see the *Olbers* room," the brothers cried and they each took one of Ali's hands. They led him away from the shadowy balcony beyond which Nature and Nature's Works were hid in night, away from the Great Illuminator who was proclaiming about the vast scope of Universal Gravitation. His companions led Ali up to a firmly closed door to one side of the museum. He noticed that the door appeared to be of exceptionally solid construction and very tightly fitted, with but the narrowest gap around its edge. This fissure was so fine as to be almost invisible, but he was amazed to observe that it was not dark. A searing thread of light, brilliant but virtually devoid of thickness, outlined the door.

"Where do you think you would be if you were to open that door?" asked Castor.

"Why, where I am now, I should think," replied Ali, slightly confused by the question.

"Not you!" Castor retorted contemptuously. "You'd be nowhere. That room portrays the golden walls of an infinite Universe. If you should find yourself in that situation, you'd be burned to a crisp upon the moment, so you would."

"I can't believe that!" exclaimed Ali. " And how about you, I should like to know."

"Ditto," said Castor.

"Ditto, ditto!" cried Pollux, then abruptly added, "Why is the sky dark at night?"

. . . .

4 I'm quoting here from Newton's own writing.

"What does that have to do with anything?" demanded Ali. "It is because the Sun is not in the sky of course. That's not much of a riddle."

"It is not a riddle," responded Pollux. "And the Sun is not important. Contrariwise! Look to the stars."

"What about the stars?" asked Ali in frustration. "They are just faint dots of light."

"Nohow! The stars are distant suns. They only look small because they are so far away. If the stars were to go on forever in space and time, then any way that you look, you will eventually be looking at a star. They would fill the whole sky, and the whole sky would be as bright as the surface of a star. The night would not be dark. Contrariwise, it would be far, far brighter than the light from one Sun in the sky."

"But whatever you say, the sky is dark at nighttime!" protested Ali.

"Indeed it is," came a new voice. Ali looked around to see a man clad in the rough durable garments of someone who explores unknown frontiers. He was carrying a heavy canvas satchel and slung over his shoulder was a large tripod-mounted tube. "We clearly see that the sky is dark when we are shaded from light of the Sun. There is nothing wrong with the argument that the sky would be bright if we could see stars uniformly spread for far enough, so we must try to find out where the stars are. That is my job. I am a surveyor for triple A maps. That is, the AAA. I have to survey the star maps for the Absolute Astronomical Atlas. The trouble is that the stars are so far away that they just look like fixed points in the sky. It is quite a tricky business to discover just how far away they are."

"I am sure it is," responded Ali politely. "How do you go about it?"

"It is not so bad for the closer stars. There we can use fairly normal surveying practices if we are careful. For those we can use the parallax method. Follow me."

With this curt command he led Ali back to the balcony overlooking the circling planets of the solar system. He did not pause by the balustrade, but grabbed Ali firmly by the elbow and before the young man had time to realize what was happening or to attempt resistance, he had leapt out into space, dragging Ali along with him. They sailed out together into the void, but before the captive youth had time to panic (much) they had landed squarely on the miniature circling blue sphere of the planet Earth, dodging neatly past the little Moon on the way in. The Surveyor planted his feet firmly on dry land, one foot on Europe and one on Africa, Ali

noted. He also became unpleasantly aware that he had himself landed up to his knees in the Atlantic Ocean.

"You will have noticed," cried his companion enthusiastically, if rather unnecessarily, "that on the Earth we are circling around the Sun. As a consequence we see the Universe from different points on our orbit throughout the year. We might expect to see the nearer stars change their position relative to the ones farther away because of parallax. It is rather as if you were trying to avoid someone in the distance by hiding behind other people. (Ali was not quite sure why he chose that illustration.) As you walk along with your gaze fixed on the person you wish to avoid, you will see that your motion will make the people behind whom you wish to screen yourself appear to move aside."

Ali peered at the distant stars as the replica Earth whirled round in its orbit. "I cannot see any movement at all," he said. "Every star seems to be fixed in its position."

"Any change in apparent position is very small, I must admit. This is because all the stars are so very far away. One of the reasons that people found it so hard to believe that the Earth orbits around the Sun was that they could not see any parallax and they could not believe that the stars would be as far away as they are. Look through my telescope and observe carefully the star in the center."

Ali peered through the eyepiece of the telescope, for such it was, that the Surveyor had unlimbered from its position over his back and pointed toward a particular star overhead. If he concentrated he could just see that the star swung to-and-fro as the miniature Earth whirled in its orbit around the Sun. The movement was really very small and was all the harder to see as the oscillation was superimposed on the star's steady drift across the field of view.

"You will have noticed that the star has its own steady *proper motion*. Stars do move through space, you know. This proper motion is superimposed on the parallax movement caused by the Earth's motion, but it is easy enough to distinguish this steady movement from the to-and-fro change caused by your orbit around the Sun.

• •

Stellar parallax

As the Earth moves around the Sun, you see the stars from slightly different positions. Most stars are so far away that there is no visible difference in the angle at which the light arrives to different parts of the orbit, but for closer stars the angle will vary.

A nearby star will appear to move against the background of the more distant stars, and the size of the *parallax angle* through which it moves allows the distance to the star to be calculated, since the size of the Earth's orbit is known.

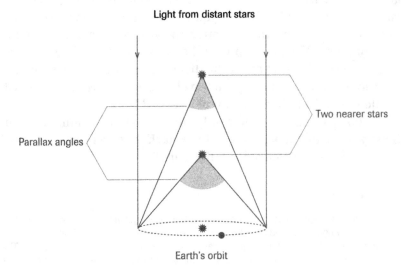

Light from distant stars

Two nearer stars

Parallax angles

Earth's orbit

• •

"As you know how far the Earth moves in half a year, then from the apparent shift in the position of a star, you can work out how far away it is," the Surveyor remarked proudly.

"How do you know how far the Earth moves?" queried Ali. This seemed to be a vital point he had missed somewhere.

"Oh that is quite easy," exclaimed the Surveyor, in the tone of someone who means 'not totally impossible.' "As the planets are all guided in their orbits by the gravitational attraction of the same Sun, their periods will depend in a calculable way on their relative distances from the Sun, and the period for a planet to circle around the Sun can readily be measured. So you see, if you should happen to know the distance from the Sun for one planet, then you can find the distance for any other.

"You may find the distance to another planet by parallax measurements, by measuring its position as seen from opposite sides of the Earth.[5] The size of the Earth is well known. The Greeks had a number for it. From the change in the observed position of the planet as seen from opposite sides of the Earth you may calculate how far away it is. When you know the distance from the Earth for several planets in the solar system, you may use Newton's celestial model to calculate their distances from the Sun. You may also find the distance to the Sun from the Earth itself. As I said, it is quite easy," he finished, still not very convincingly.

"Anyhow, the distance to the Sun from the Earth is a vital quantity in astronomy. It is called the astronomical unit, or AU for short. As I was saying, when you can measure a parallax movement for a star, you can work out how far away it is. It turns out to be a long way. We come up with distances of tens, even hundreds of light years. A light year is the distance that light will travel in a year, and that is a long way. To travel all the way around the Earth would only take light a fraction of a second."

"That's amazing," said Ali. He felt it was pretty amazing, but mostly it seemed to be the sort of remark that was required. "And can you find the distance to all the stars this way?"

* * * * * * * *

5 You may also get a very good measurement of the distance from the Earth to another planet by bouncing a pulse of laser light from it. As you know the speed of light, then measuring the time it takes light to travel there and back gives you the distance. This process has been used to measure the distance to Venus, and the distance to the moon has been measured at various times to high precision by installations like the satellite ranging observatory on Maui.

"I am afraid not. The farther away a star is the less apparent shift you get as the Earth moves. For distant stars the motion is too small for your telescope to reveal."

"Couldn't you just increase the magnification of your telescope, then?" asked Ali. It seemed a fair suggestion.

"I am afraid that it would not do you any good. The image of a star in the telescope is not only small, it is also slightly blurred. If you had greater magnification you would just enlarge the blur and learn no more. It is all because of *diffraction*. Let me illustrate that.

"Hold your hand up in front of a bright background so that your thumb and forefinger are almost touching," commanded the Surveyor.

Ali did as directed, holding his hand up to the light of the little sun and bringing his thumb and forefinger steadily together. Just before they touched he saw a blurred dark band that appeared to bridge the gap between them[6].

"You will have observed a dark region appear in the narrow gap. This is an effect of diffraction. Usually, light appears to travel in straight lines because the light waves come in on such a broad front that only those going straight ahead will interfere constructively at their destination. Away from this forward path, light from different positions along the wave front will arrive out of phase with one another and cancel out, in the process of destructive interference. With a narrow aperture, however, there is no broad front to give all the canceling waves. You have seen that you got a fuzzy blur where you might have expected a narrow line of light between your fingers. This is diffraction."

The Surveyor indicated his portable brass telescope. "If you have a small telescope you will see a fuzzy dot where you might expect a tiny sharp image of a distant star. If you have a large telescope with a wide aperture you have a wider wave front entering the telescope and narrower spread in the light due to diffraction. This means you can get a much sharper picture with a large telescope.

"Nevertheless, however large you make a telescope you still cannot make the image too sharp since when you look at the stars from the surface of the Earth, the light has to pass through the air above you and that air shakes and flickers with thermal turbu-

• • • • • • • • •

6 You can try this at home.

lence and haze. The light from the star is spread as different parts of the wave front travel along different paths through the air. As the air is never still, the light is deflected differently by different regions through which it passes, and you see the star twinkling as the atmosphere shimmers above you. As the rhyme goes:

Twinkle, twinkle little star,
What a fuzzy thing you are.
Up above the sky you fly,
This blurs your image to my eye.

"I'm sure that I remember something quite different," protested Ali.

"Never mind. The fact remains that you cannot get a precise enough image of a star to measure its parallax if it is more than a couple of hundred light years away." The Surveyor looked rather saddened by this, but rallied quickly. "Oh, there are ways of making estimates of the distance from the Earth to *groups* of stars, and these work for greater distances, but they are all limited"

"And are there then many stars that are so far away?"

"Oh yes! Almost all of them are, in fact. Billions of billions of stars are too far away for any practicable measurement of their parallax. Most of them are too far even to *see* without a powerful telescope. So how can we tell anything much about the stars as we look at them from the surface of the Earth, below the sky and often below clouds, below an atmosphere that distorts our seeing? Let us go below the atmosphere ourselves."

The Surveyor reached out and again seized Ali's elbow. Together they began to shrink toward the surface of the miniature planet that became less miniature by the moment. As they descended, he pulled Ali toward him, hauling him out of the water and guiding them both down to the same portion of the planetary surface. In short time they found themselves standing on a sparse desert floor, looking up at a tall rocky pinnacle that was topped with a heavy squat building beneath a pale dome. Above all rose a black night sky thickly sugared with stars.

"That is an observatory dome–home to a massive telescope much larger than my portable one. It is a great tube containing a curved mirror several meters in diameter that serves to focus the light from a distant star down to a single point. More or less," he added. "Now you might ask why you want such a large mirror. What is wrong with my small portable telescope?" he asked rhetorically. Ali rightly guessed that he was not supposed to answer and simply waited.

"I have told you the first reason. It is because of diffraction at the telescope's aperture. The second reason is quite straightforward. More light will fall on a big mirror than a little one and so you can hope to see much fainter stars." That sounded logical enough.

"To find out more about stars you have to look at them more carefully. You must observe *everything* that you can see from our restricted observation point, anchored to this single Earth. We must look at the stars—not just glance at them as a distant vista that fills the night sky, but *really* look at them. We must examine them in all detail that is available to use. Just look at them," he commanded. "Do they all look the same to you?"

"Well," responded Ali judiciously, " some look to be a lot brighter than the others."

"Exactly! The brightness of stars is one measurement we may make. Come up and meet a colleague of mine who is an expert on measurement."

They labored up a steep and winding track that led eventually to the door of the domed observatory building. The door creaked open to reveal a great hunched figure that loomed over them, straight out of a traditional horror story. This gnarled and hairy shape was complemented by the possession of a truly enormous nose. Balanced on this was a pair of heavy thick-framed spectacles through which peered a pair of surprisingly mild nearsighted eyes. At least he was wearing the spectacles *most* of the time. From the spectacles hung a loop of cord whose purpose was clear as Ali noted that the spectacles were so heavy they would keep slipping down his prodigious nose, and from time to time threatened to drop off completely. To prevent this, he draped the cord around the back of his neck, so if they fell, he could catch them readily enough.

"Come in," the nightmare shape invited in an inappropriately mild and friendly voice. "Come and see my telescope. I'm rather proud of it."

They followed him as he shuffled into the dome, where they discovered an elaborate framework supporting the squat tube of a great telescope. There were mysterious boxes and cables draped here and there over it, but no obvious place for anyone to look through it. Ali commented on this.

"Oh, I never actually *look* through it. Dear me, no, I am not very good at actually looking at things," he remarked, adjusting his spectacles. "I am a Techno Troll and all my observations are made by instruments that give me a numerical readout. Observations must be quantitative if we are to make a map of the heavens that is accurate," he said emphatically. "They say that seeing is believing, but measuring is *knowing*. To make estimates of quantities like the distance to a star, we need numbers. A value for distance is a number and if we are to finish with numbers we must put numbers in to start with. Just looking through a telescope by eye is not sufficiently quantitative," he remarked, turning on them his weak protruding eyes set below great bushy brows. "Eyes are all very well for many purposes, of course, like moving around without bumping into things, ha ha!" The Troll laughed at his own whimsicality. This caused his ungainly body to shake so much that his spectacles dropped off again. He turned around as he was fumbling to put them on and fell over a chair.

"The Surveyor said that you had to look at the brightness of stars," put in Ali. He had no idea why this should help, but he wanted to cover the Troll's embarrassment.

"Yes indeed, we can readily measure the amount of light that we get from different stars," he enthused as he scrambled to his feet. "This amount is commonly expressed as the *magnitude* of the star, a brighter star having a smaller magnitude. A really bright star would have negative magnitude. It is all done with logarithms and other mystical operations," he added somewhat uninformatively. "The brilliance of a given star will depend both on how bright the star actually is, which is its *absolute magnitude*, and on how far away it is. The further away a star, the fainter it will appear. The light that a star emits will shine out in all directions so at any distance it will be spread across an imaginary spherical surface centered on the star. The further away the star then the larger is this sphere and the smaller the fraction of its light that will enter your eye or your telescope. As I was saying, it is hard to tell how bright a star appears by eye. We used to do it by taking photographs, but this is not very efficient. A very long exposure was needed, often of many hours, and the darkening on the film negative was not easily related to the amount of light. Now we use electronic arrays that are much more efficient, about ninety percent of the light being detected, and the signal we get is proportional to the amount of light. It is really much more convenient, another triumph for technology," said the Techno Troll.

"It is hard to tell just by looking at it whether a star is near and dim or far away and very bright. If you see a brief flicker in the night it is hard to be sure whether it is a firefly in your garden or a lighthouse on a distant shore. If, however, you should happen to know how bright the star *really* is, then from the amount of light that reaches you, you might discover how far away it is. A candle near your elbow will give enough light for you to read by, but one at the other side of your village will be a barely perceived glimmer in the night, and once you know that what you are seeing *is* a candle, you may estimate its distance."

• •

Stellar magnitudes

Some stars appear much brighter than others. This brightness is traditionally expressed as a magnitude. The Greeks divided the stars into six magnitudes, the first being the brightest and the sixth the faintest that could be seen. The description is now more precise but it remains true that the smaller the magnitude the brighter the star. An increase of five in magnitude means the star is actually one hundred times fainter.

Apparent magnitude is a measure of how much light reaches the Earth from a star. This, of course, depends on the distance to the star as well as how bright the star actually is.

Absolute magnitude is the magnitude a star would have if it were 10 parsecs from the Earth, one parsec being 3.26 light years. Absolute magnitudes give a comparison of actual light emission from different stars.

• •

"That sounds fair enough," admitted Ali. "So how can you tell how bright the star really is?"

"Well, that is the problem of course. Stars are by no means all the same and certainly they are not all equally bright; they come in many types. Some are spendthrifts who squander their substance in a short time with a huge display of brilliance. Some have but a meager horde of fuel that they use as misers would—a weak candle that they can eke out for billions of years while sitting almost unnoticed in the night. Fortunately we can recognize the various types from the nature of the light they send us. The brightness we see on Earth will not identify them because we do not know their distance, but their color can."

"But stars are all white!" protested Ali.

"Oh, they are not! They may seem so to your eyes, but not to my instrumentation," said the Techno Troll proudly. "Because the image they produce on your eye is so small, you cannot see their color, but color is there. My instruments can see it clearly. Look at stars with a large telescope that will gather enough light and you see them colored as they are. Look at them with the appropriate instrument, with a spectroscope, and you can see how color is distributed within their light. Come!"

At the floor level of the dome there was a door. The Troll opened this and they trooped into what was obviously a control room, pausing only briefly while he bumped into another chair. Inside were banks of panels with dials and switches and pretty colored lights. Scattered around were screens like small TV sets and chairs modified for the Troll's rather unusual physique. Displayed on many of the screens were stars. Some screens showed great seas of stars. Some were centered on a single star, though there were usually one or two others on the periphery of the scene. Some showed strange graphs and plots whose meaning was not immediately obvious. Behind it all they had a sense of endless electronic activity. Their host led them to one such screen on which was shown a band of color, varying from a deep red on one side through orange, yellow, and green to blue and finally a deep violet, beyond which the band faded from sight. He indicated the spread of colors on the display with a wave of a great gnarled hand.

"This is a spectrum. The light that entered the telescope from a star has passed through a spectrometer and is now separated into light of different frequencies. This is much like the way that raindrops can separate sunlight into a rainbow of colors.[7]

This spectrum has been recorded electronically in my instruments and we know the intensity of the different colors in the light. We have the numbers and from the numbers we can learn many things about the star. One useful fact is the temperature of the sur-

. . . .

7   The first and simplest example of a device to create a spectrum was the glass prism used by Sir Isaac Newton to split sunlight into a rainbow of colors.

face from which the light has come. The colors are distributed as you see here."

He leaned over the working surface before the screen to touch a switch, knocked over a cup of coffee in the process, recovered his spectacles, which had slipped down the long incline of his nose, and continued. "This is a typical spectrum for the light that you get from a hot body, the spread of color that you get from a hot glowing furnace or gas. There is a continuous spread of color and this reaches its brightest at a frequency that depends on the temperature. You may equally well talk of frequency or of wavelength for the light," he added by way of explanation. "A wavelength is the length of one oscillation of the light and frequency is the number of oscillations that you get per second. The product of the two, the length of an oscillation and the number that pass per second, is simply the speed at which the light travels, and this never varies. As speed of light is known and unchanging, then if you know either frequency of wavelength, you may quickly find the other. A long wavelength thus means a low frequency, and vice versa. Near normal room temperature the peak of brightness is in the *infrared.* This is light of a wavelength even longer that the red light you are able to see."

"How can you have light that you cannot see?" asked Ali. "Surely the whole point of light is that you can see it?"

"Perhaps I should not have said 'light' then, but rather electromagnetic radiation: a traveling system of oscillating electric and magnetic fields. 'Light' is a rather parochial term. If the frequency of the oscillation falls within a certain range, then you have light that your eyes are able to see: visible light. The possible frequencies spread far on either side of this range. Frequencies higher than that of blue light result in ultraviolet light and higher yet give X-rays and gamma rays. Frequencies lower than found in red light give infrared radiation and yet lower you get microwaves and radio. All frequencies are not only possible in principle, but they are all found in nature. The wide range of frequencies is shown in this plot. Colored bands down the plot mark those in the range that your eyes may see. Let us look at the spectrum for a very cool star."

He summoned up a plot in which the humped curve of the radiation spectrum fell almost entirely below the visible region. "This would be a very dull red star. A much hotter one will be yellow or even blue-white." The spectral curve now stretched up to higher frequencies so that its peak was above the blue end of the visible spectrum.

"You can see how color and temperature go together if you withdraw an iron sword blade from an intense furnace. At first it blazes with a white heat, but as it cools, this fades through yellow and orange to a dull red. Even when it goes totally dark to your sight it will still be glowing in infrared and could give you a nasty burn.

• •

Thermal radiation spectrum

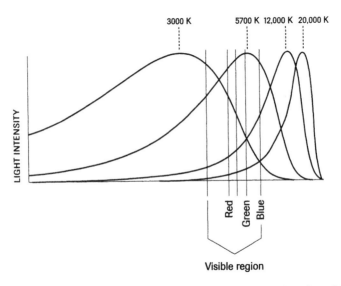

The continuous spectrum of thermal radiation emitted by a hot object moves to higher (bluer) frequencies as the temperature increases. At 5700 degrees Kelvin[8] a lot of the radiation is in the visible region. This is roughly the temperature of the visible surface of the Sun.

• •

"In much the same way we can recognize hotter and cooler stars by their color. We can compare the amount of light that stars of different color and temperature actually emit by looking at the

• • • • • • • •

8 A degree Kelvin is the same *size* as a degree centigrade, but it is measured from absolute zero.

brilliance of different stars in a distant cluster. As these stars are all much the same distance away, their absolute light outputs will be in the same ratios as their measured intensities. We may plot this against their color."

He placed his spectacles firmly back on his nose and reached out a long arm to press another control. The screen now showed a diagonally sloping band of dots. "This shows the intensity of stars and their color. You will observe what is called the *main sequence* in which the stars follow a definite relation between brightness and color. At one end you have stars that are red and very dim, at the other end are brilliant white stars that give out a million times more light than our Sun.

"Now, we know the distance to some of the nearer stars from various forms of surveying." He gave a nod toward the Surveyor as he spoke. This turned out to be unwise as his spectacles promptly fell off again. "We can observe the apparent brightness of these, and with their known distance from us we may find their absolute magnitudes, the total amount of light they emit. From the spectrum of their light we know their temperatures and can see where they fit on our plot. In this way we can calibrate our chart and show how color relates to absolute magnitude. From this you would expect that you could tell the amount of light a star of a given color really emits and, from the fraction of the light that you actually observe, you could tell how far away the star is.

"You could, that is, if every star lay on the main sequence. Unfortunately, they don't. There are some rather glaring exceptions, if you will pardon the pun." He coughed gently into an enormous hand. "There are red giants that emit enormous amounts of light and white dwarfs that are rather dim, so color is not after all quite enough. What you need," he went on, removing his spectacles and polishing them carefully, "is an absolute beacon. You need a type of star that always gives the same amount of light and that may reliably be recognized in distant clusters. The distance to such a standard star would be easy to estimate from the amount of its light that reaches the Earth, and a number of such stars would give a grid of fixed points that would in turn position the stars around them.

. . . .

Main sequence stars

If you make a plot of the brightness of a star against its surface temperature or color, you find that they fall mostly on a band known as the main sequence. In general the cooler (red) stars are fainter and the hotter (blue-white) stars are much brighter.

There are side branches for the very bright red giants and the dim white dwarfs.

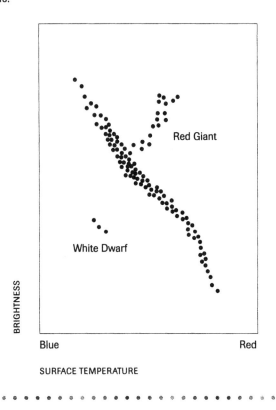

. . . .

"We do have such an identifiable standard. There are stars whose true output of light we may know however far away they may be. We have found our constant in inconstancy. These markers are Cepheid variable stars: Stars whose intensity varies predictably over a period. They provide a flashing beacon that guides us to the scale of the Universe. It turns out that the longer the time it takes for the intensity to fade and rise, then the brighter is the star. The two are accurately related. The relation between the time over which one of these stars fluctuates and the amount of light it emits was mapped for populations of stars in distant clusters. These stars are all much the same distance from the Earth, so their relative brightness as observed will be proportion to their absolute brightness.

"All that remained was to find the absolute brightness of a few examples of these variable stars and then the absolute brightness of a Cepheid variable anywhere in the Universe, can be found from the relation just established. The distance to a nearby variable star of this type was found by methods akin to parallax. Once you know how far away a star is, you can readily calculate its absolute brightness for the amount of light you detect from it.

"Armed with all this information, whenever we recognize one of these variable stars, we know how bright it must be and can hence find its distance. This works well for as far as we may see and recognize these stars. As you look to greater and greater distances, there are many methods that may give a scale, but it does get complicated," he admitted.

"There is much more that may be learned by looking at the light from stars. You normally learn about your surroundings by looking, which means looking at the light from them. This remains true even for the tiny dots of light from stars. The light can tell you what a star is made of and even how fast it is moving away from you."

"How can it possibly do that?" demanded Ali. "It is only light, just a twinkle in the sky."

"Come and see," commanded their host. He led them back to the screen on which they had previously been shown the colored spectrum of starlight. "Look more closely," he commanded. Ali peered at the colored band and noticed a great number of black lines across it: a myriad of extraordinarily thin slices where the color had been removed. "Those are absorption lines," he was told. "Every atom has its own distinct set of frequencies at which it will emit or absorb light.[9] Here you see the lines that atoms around the hot furnace of a star have left when they absorbed the light at their own distinctive frequencies. Because this set of frequencies is unique to all atoms of a given type, you can tell what chemical elements are in the star,

. . . .

9 See the Snow White tale. The atomic electrons must occupy levels of distinct energies and as they transfer from one level to another, they absorb or emit light of definite and identifiable frequency.

even though it is so very far away. The gas helium was discovered in the Sun before any was found on the Earth. It is from such spectra that you can tell what stars are made of.

"Perhaps surprisingly, this same light can also tell you how a star is moving. When you look at the spectral lines from a star and compare their frequencies with light emitted by the same elements in your own laboratory, you will see the same *patterns* of lines, but often these line are all displaced slightly in frequency. This is an effect of the Doppler shift,[10] and from the amount of shift you may estimate how fast the star is moving toward or away from you. At great distances we find that whole galaxies of stars are moving steadily away from us. The further away the galaxy, then the greater is its speed away from us. This breakneck dispersion of the galaxies is known as the 'Hubble expansion' of the Universe.

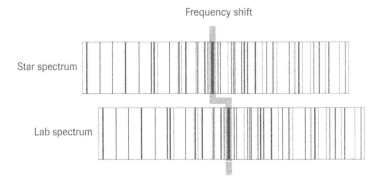

Doppler shift velocity

Starlight contains clearly recognizable spectral lines - unique signatures of the elements in the cooler outer layers of the star. These absorb specific frequencies from the continuous spectrum of light that comes from the hot interior. In the light from more remote stars, these lines are often displaced in frequency, usually towards lower frequencies. This *red shift* is from the Doppler effect and gives a measure of the speed with which the star is moving away from us.

For some of the more remote stars the shift is so great that familiar lines move from the visible region into the infrared. In some cases they move so far that when they were first observed, the spectra were not recognized.

"The most remote galaxies, ones so far away that only the largest telescopes may see them at all, are receding from us with speeds that are beginning to approach the speed of light. The whole Universe is expanding, so as time goes on it is steadily becoming ever larger. Conversely, we may assume that as you go farther into the past, the Universe was ever smaller and, as you extrapolate back from this, there would be a time in the past when it had no size at all. From that time to the present may then be understood as the age of the Universe, and that age is *finite*. The Universe apparently has not been around forever, though by our standards it is pretty old: billions of years."

"You speak of huge distances. Is the Universe indeed so very large? Such dimensions are beyond my comprehension," remarked Ali.

"Oh yes, the Universe is vast," the Surveyor answered. "There are stars so distant that light from them has taken billions of years to reach us. Come outside and we can check this out in a recent edition of the *AAA*.[11]

The Surveyor turned to go and their grotesque but nonetheless gentlemanly host gave a little bow to acknowledge their departure. This, of course, resulted in his spectacles falling off and this time they slipped off their cord and fell to the floor. Despite this, he courteously attempted to escort his guests to the door. Ali and the Surveyor managed to avoid his polite intent to usher them into what in fact appeared to be a broom closet, and made their own way to the exit.

Outside they descended the winding path to the desert floor, and when they reached the ground the Surveyor opened his faithful satchel and took out a very large book with the letters "AAA" stamped on its spine. "It is difficult to appreciate directly the huge numbers that one feels moved to quote when speaking of the size of the Universe; they are numbers of the sort that is usually called 'astronomical.' It is probably better to look at a sequence of comparative sizes that are not too extreme, and by this means approach the vastness of the Universe in a sequence of fairly easy steps." He opened the book to a page that showed a lone figure

· · · · · · · · ·

10 See The First Tale, "The Prince and p." Light from a moving object is shifted in frequency.
11 That is the Absolute Astronomical Atlas, if you remember. I need hardly add that this publication is purely imaginary, though the information presented is genuine.

standing in a desert. Ali recognized himself; it was as if he were looking in a mirror.

"There you have the most familiar part of the Universe–yourself. If you want to consider the Universe overall, you must obviously look on a broader scale. It is difficult to imagine huge numbers, but you can probably visualize something that is a hundred times as large as you. That difference involves a variation in sizes that your eye may readily comprehend; indeed, it is roughly the difference between your own height and the height of your eye. It is not at all what you might call an astronomical number. Each successive page here has been drawn to show a region that extends one hundred times as far as the one before."

The Surveyor turned the page to reveal another picture with the same figure in the center. It was tiny now but still visible, and near it the whole of the path that led up to the observatory could be seen. Another page and the picture showed a wide expanse of desert surrounding the rocky peak that supported the observatory. Nearby was an oasis with a luxurious growth of palms and other vegetation and, away to one side, a row of pyramids. The next page showed a wide, dry desert land from whose coast stretched an expanse of ocean, right up to another continent on its far shore. The distance showed a horizon that looked distinctly curved.

Things changed with the next page. It was filled with the black of space apart from a blue and white sphere that occupied less that a twentieth of its width and, right at the edge, a silver-gray sphere that carried the familiar cratered markings of the Moon. The Atlas was now clearly astronomical.

Following this, the next view revealed more planets, pale dots that were the inner core of the solar system. All attention now focused on a fierce blaze in the center. That was the first star to appear: our own Sun whose fiery disk dominated the scene.

With little pause the Surveyor turned over another page. Again the black of space filled most of the page. This publication must use a lot of black ink, thought Ali. The sun was still the most obvious feature, but now it had shrunk to a small though brilliant disk in the center. Tiny dots showed the positions of the inner planets, and a few small disks with rings marked gas giants like Jupiter and Neptune. Saturn was especially splendid with its dense ring structure.

The next expansion by one hundred saw the Sun appearing as a bright star, with most of the planets too small to be noticed. A vague and almost imperceptible dust of tiny shapes around the region occupied by the solar system marked the Oort cloud, the region occupied by a swarm of small bodies and from which they occasionally dived in close to the Sun as comets.

The Surveyor turned over yet another page. Now the Sun was visible only as a rather faint star. To keep it company, another star had appeared toward the edge of the page.

"Now we have come to interstellar distances. Even the nearest star is a long way away, some thirty million million kilometers, but we have reached it with only eight enlargements from your own size. In each of these we have changed by the comparatively small factor of one hundred. We have a few stages still to undergo before we reach the edge of the visible Universe." He turned a page again and now the scene was full of stars. Faint ones, bright ones, blue and yellow and red ones. There appeared to be millions of them and it was quite impossible to identify the Sun amongst them all.

Another turn of the page, another factor of one hundred, and the stars had blended into a milky haze. The accumulation of stars was spread out across a great disk with twisting spiral arms and a central bulge where the blaze of starlight was more intense.

"That is our galaxy. It contains hundreds of billions of stars. With its central bulge surrounded by a wide disk, it is like a great child's top slowly spinning through space, and you see that, like the top, it is decorated with bright radial bands of more intense activity that spiral out from the center. Our galaxy is by no means unique. If you look further out," he remarked as he turned another page, "you will see more galaxies that form a *local cluster*. Beyond this (another page) you may see that the clusters of galaxies themselves congregate into threads and knots that meander through space, to form a sort of sponge of galaxies. Separating the strings of galaxy clusters are great empty voids. And finally," he turned another page with an appropriate air of finality, "you may see all of the visible Universe represented." The picture showed a great web of dispersed stellar matter separated by great voids and all fading out toward the edges of a circular region. Beyond this the page was completely blank. "The scale of the whole Universe that you may see can be portrayed in these thirteen steps of moderate size."

"What lies beyond these pages? Is that last region shown, vast as it may be, indeed the whole size of the Universe? Does the Universe have some sort of end or boundary at that point?" asked Ali, who had become quite fascinated by the steady advance shown in the successive pages.

The Surveyor turned over a couple more pages. They were almost entirely blank. "Who knows what lies beyond?" he said. "Our Atlas shows the size of the *visible* Universe and that is all that we can see and measure. As far as we know there is no 'edge of the Universe,' no boundary in space. The Universe *has* a boundary, though, and that is a boundary in time. You must remember that the galaxies we see are long ago and far, far away. Because it takes the light so long to reach us from a galaxy that is far away, we can only see it as it was long ago. As we look far out, we are looking far into the past and eventually we come to the birth of the Universe. The farthest position at which you may see any stars is not as far away as the present distance to those stars, because the Universe has expanded further since the light set out toward you. It is, however, as far as you can actually *see*. On a clear day you cannot see forever." The Surveyor slammed his book closed with an air of finality, put it back in his satchel, and together he and Ali set off across the desert.

They had not traveled far when they came to a figure sitting cross-legged by the roadside. He was swathed in robes and had a strip of cloth across his eyes. "Probably a blind beggar," said the Surveyor, preparing to walk on past.

"No, it is you who are blind," the shrouded form contradicted him. Rising smoothly to his full height he stared at them through his cloth bandage with eyes that seemed to see perfectly even though they were covered. "I, the Broadband Visionary, say that you who look at the sky with your optical telescopes see the Universe through a veil, the veil of the Earth's atmosphere. It has a few slits in it. One of these allows a narrow band of light frequencies to reach the ground: the range that you call visible light."

"Well," responded the Surveyor, "visible light is all we can see by anyway, so that is not much of a problem."

"Oh but it is a problem. It is not too surprising that people have developed eyes that use the light that is available to you on the Earth's surface to see by, but the range of electromagnetic radiation is much more than that. The electromagnetic spectrum extends over a huge range of frequencies that you cannot see, and these can tell you a great deal more about the Universe around you. Frequencies to either side of those

that your eyes see, in the infrared and ultraviolet, are indeed blocked by the atmosphere, but much lower and higher ones can get through. Radio and gamma rays from space do reach the surface of the Earth. By observing other frequencies of electromagnetic radiation you see marvels not visible to your eyes. If you restrict yourself to that tiny band of radiation that your own unaided eyes may detect, you make yourself deliberately blind."

Atmospheric absorption

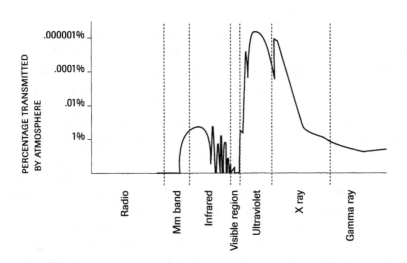

The Earth's atmosphere is largely opaque to many frequencies of radiation. There is a narrow window that passes the range of frequencies our eyes can see, and a broad band of radio frequencies reach the Earth's surface easily. Other frequencies of radiation are blocked to a greater or lesser degree and are best detected by satellite instruments that are above the atmosphere.

"What use to us are other radiations that may get through?" returned Ali. "You cannot see radio waves, can you?"

"Well, actually you can. Or your instruments can. Radio waves will induce tiny electrical currents in a piece of wire, a radio antenna. If you collect and focus such radio waves, you can make an 'eye' that is able to see radio. Behold!"

He rose to his full height and pointed a long cloth-draped arm toward the expanse of desert. "Look near where those vast and trunkless legs of stone stand in the desert, there beside the shattered visage lying on the ground. Look where the lone and level sands stretch far away." They looked where the desert floor stretched on beyond the shattered colossus he had indicated, and as they watched, the ground began to heave up and up. As the mound continued to rise, rivers of sand streamed down its sides to disclose a wide shiny black hump. This struggled free of the sand to be revealed as an awesome beetle-shape on a gargantuan scale, poised on six legs on the level desert. It dwarfed even the ruins of the great statue close by.

"What is that?" whispered Ali.

"It is a microwave scarab, a detector for the radio waves that whisper to us from space. See!" commanded the Broadband Visionary. Slowly and impressively, the wing cases of the great beetle shape unfolded and rose above its back, swiveling round as they extended to form a shallow circular dish. The creature, if it was a creature, lifted its antennae to lie at the focus of this reflecting dish and shuffled around on its several feet to point the bowl into the sky.

"There you have the microwave collecting dish of a typical radio telescope. Well, perhaps this one is not entirely typical, since few are mounted on the backs of giant beetles. The dish shape that focuses the radio waves onto a detecting antenna is common enough. The parabolic shape will focus waves only from one direction onto the antennae. The sharpness of the focus will be limited by diffraction as with an optical telescope. The resolution

of a small telescope is restricted by the width of its focusing mirror and the same relation applies here. The bowl here may be over a hundred meters across, but this is still a small aperture for radio focusing. The frequency of the radio waves you would wish to see is very low, its wavelength is hundreds of thousands of times greater than that of visible light, so that great dish cannot see as sharply as the unaided eye."

"I don't see a lot of point in it, then," grumbled the Surveyor, who was irked by the challenge to his lifetime of observations.

"Oh, but there is," countered the shrouded Visionary. "Different frequencies will reveal different things. You know that objects with the temperature of stars emit visible light copiously, but there are other things in the Universe that are much cooler, and indeed some that are hotter. These are not seen in visible light but may show up when viewed with a different frequency of radiation. Radio waves are freely emitted by much cooler sources, and radio allows you to see great clouds of gas heated slightly by nearby stars.

"A radio dish such as you see before you will focus on a single point in the sky, but if it is scanned around it will build up a picture of a whole region. You may take your time to build up a picture as most features in the sky change only very slowly." They watched for a moment as the great shape shuffled about on its angular legs, moving steadily from one point to another.

"You can get a picture of radio sources in the sky, and they are dramatically different from the distribution of stars that you see in visible light. You see clouds of gas that block out your view of most of this galaxy. If you look further afield you may see dense clouds of gas within which stars are being born, and fast-moving jets shot out from stars as they die. In order to see such distant features, you must have better resolution, so you need a telescope with a much wider aperture."

"Isn't that monster already big enough for anything that you could require?" protested Ali.

"No, by no means. As I have just said, the radio waves it uses are much longer than the wavelength of the light by which you normally see, and so the width of your telescope most be proportionately larger. I have told you that this dish before you is unable to match the resolution of the unaided eye. To approach the precision of a large optical telescopes, you would need a receiver that is much wider, something hundreds of kilometers across."

"Well, that is hardly practicable!" snorted the Surveyor.

"Ah, but the detector does not have to be solid or continuous. It does not even all have to be there at the same time. You require only a few dishes scattered over that sort of area. With radio waves you have the advantage that you record the phase of the waves as you detect them, and so you may at your leisure combine together signals that you record as the Earth carries your telescope through the incoming waves. In this way you may recover the same precision that you might get all at once from a single huge receiver. See what I mean."

Away beyond and to the side of the great detector dish, the desert floor was erupting. More vast dark shapes crawled from the sand and spread their dishes to eavesdrop on the sky. Small spider-like shapes emerged and scuttled between each one, trailing thin strands of cable to link up their signals. Soon there was sitting on the sand an extensive array of dishes, aligned so that they could combine the signals received from the same point in the sky.

"There you have a high precision radio telescope. We call it the VLBA, or Very Large Beetle Array.[12] This array of insectile dishes is the equal in resolution of any optical telescope on Earth. It is superior in fact, because the radio waves do not suffer from the same degree of scattering from the atmosphere. You may not be able to see radio waves, but the 'seeing'—that is to say the disturbance produced by the atmosphere—is much better.

"Other frequencies give you other information. You see a different view in infrared or ultraviolet or X-ray frequencies. For these frequencies you cannot site your telescope upon the surface of the Earth because the air above will block the radiation quite effectively. Your instruments must be raised into space." He gestured toward some distant insect-like shapes with great compound eyes of tubes, mirrors and lenses. As they watched, the shapes crawled up the side of what looked like a tall pillar and shortly thereafter flames erupted beneath the pillar and it rose skyward, riding on a column of fire. In due course it would be surveying the heavens from above the Earth's blanket of air and radioing back its observations. As they looked around more carefully, they saw yet other instruments scattered around the landscape. One great array of glass lenses looked rather like a gigantic fly's eye[13] as it peered

· · · · · · · · ·

12 The VLBA does exist as a great array of radio detecting dishes, but the name actually stands for Very Long Baseline Array. The principle is the same as for this rather spurious version but as it happens its dishes are not mounted on giant scarab beetles.

13 There is a detector called the "Fly's Eye." It is a huge array of phototubes in the American desert that tracks very high-energy particles by light they emit when they interact with the atmosphere.

up at the sky, waiting to catch the light emitted by cosmic particles of inordinately high energy as they blazed down through the atmosphere.

Another object that caught their eyes was a large angular cone, a sort of squared off "horn of plenty" that was supported by an adjustable mounting on the desert floor. They asked the Visionary what was the purpose of this. "That is a microwave horn antenna. Just such a one was the first to detect the Cosmic Background Radiation."

Ali was impressed by the capital letters that the Visionary managed to attach to this phenomenon, even in his speech, and so asked to know more about it.

"If you look at the sky with a detector that will register radiation of lower frequency, and thus of lower energy, than the waves recorded by the arrays of radio dishes, then you detect a signal. You find that, unlike most of the signals from the sky, it does not come from anywhere in particular. As you point the detector in different directions, it makes no difference to the strength of the signal you receive. It appears to come from nowhere and to be everywhere. When you look at how the intensity varies with different frequencies you find that, like the light from stars, it follows exactly the form of radiation from a hot body, but in this case the hot body is very cold. Rather than coming from an object at a temperature of thousands of degrees, it is appropriate for one less than three degrees above absolute zero. You are seeing the fading whimper of a great fireball that filled the whole Universe billions of years ago.

"Now, you may say that seeing is believing, but I hope I have shown that you would be ill-advised to limit your view of the heavens to just what your eyes can see, even when they are boosted by the mirrors and lenses of a telescope. What is so special about wavelengths between two thousand and six thousand angstroms, after all?[14] You miss a lot if you do not take a wider view. Your view of the Universe should not depend only on what your eye can see, as your unaided eye can see so very little of it. Rather, you must consider all that your observations reveal, however they are made. Your picture of the Universe will include not just what you your-

• • • • • • • • •

14 This is a traditional measure of wavelength. One Angstrom is one ten billionth of a meter.

self can see but what your mind has discovered. Come and see the Universe as seen by the Mind's Eye."

The Visionary set off determinedly across the sand and they followed him to a low cliff and into a cave in its side. Together they strode down a rocky corridor and into a wide, arched room. At the far end was an immense head, apparently carved into the rock. One of the visage's eyes was closed in deep inward contemplation, but the other stared wide open. A curving flight of steps led up to this eye, and they climbed them to stand on a stone platform just in front of the great eyeball. The eyelids and iris were all carved of stone, as was the rest of the head, but the pupil was a strangely sinister dark pit that appeared to open into infinity. They stood shoulder-to-shoulder and peered within at its unthinkable depths.

"They say that the eye is the mirror of the soul, but in this case it is the window of the Mind. Looking within you may watch as the Mind contemplates the Universe beyond. These visions will go far beyond what your eyes can see. The Mind broods upon a picture of the Universe based on the best information revealed to present knowledge."

As they peered into the interior perceptions of this remote inhuman mentality, they saw at first a view not too different from the desert sky as they might see it themselves at daytime. The fiery disk of the Sun dominated the scene, but here it blazed closer and more ferociously than they had ever seen it from the Earth. It was much too bright to look at directly, vast within their field of view. A little sphere swung by and lost itself against the blaze of the Sun. "Mercury," breathed the Surveyor near Ali's ear, "making a transit of the Sun."

The disk of the Sun shrank as the Mind changed its focus and their viewpoint drew back and back and back. Planet after planet was presented to their view as in turn they saw the Sun from the vicinity of each. There was Venus and the familiar Earth and little red Mars. Then, after a clutter of small irregular asteroids, there were the gas giants: planets with no clearly solid surface. Massive Jupiter appeared with its prominent red spot, a hurricane larger than the Earth that had persisted throughout human history. Next was Saturn with its fantastic set of rings, but otherwise a smaller version of Jupiter. Neptune followed and Uranus and then Pluto, the stunted orphan of the solar system. From this remote frozen planet the Sun appeared little more than an unusually brilliant star in the sky.

Abruptly the Mind's capricious focus shifted to another view-point entirely. A dense field of stars shrank as the viewpoint drew back to reveal a circular wheel of stars: our own galaxy. A milky mist of stars, so many that from this distance they merged into a hazy cloud. Billions upon billions of them made up this slowly turning entity. It had a great central bulge in which the stars shone mostly blue white, whereas those in the disk were dominantly yellow and red. Around the disk were curving radial spokes that spiraled out from the center and glowed in places with the light of blue-white stars: new stars born from denser regions of gas within the churning arms. Scattered through the galaxy were regions of cooler dust and gas that veiled the central bulge from Earth's view.

Around the Galaxy were strewn *globular clusters*, small com-panions to the main collection, though each was composed of hun-dreds of thousands of stars. The onlookers were somehow aware of a great mass of dark matter that spread within and around the disk of the galaxy, permeating it and extending far beyond. This was not dark like the dark clouds of dust that glowed gently with microwave radiation. This matter was truly *dark*, emitting no detectable radiation at any frequency. It manifested itself only by its gravity, which affected the motion of the stars in the galaxy and held them in orbits around the galactic center at positions and speeds that the mass of visible stars and dust could not account for.

It was evident from this contemplative view that apart from one or two local exceptions, all of the galaxies were rushing away from one another. The farther they were from the Earth, the faster they appeared to move. In fact, their velocity of retreat was in proportion to their remoteness, apart from minor variations that indicated galaxies had their own random motions superimposed on the overall grand retreat. The galaxies were borne apart by an expansion of the space that contained them. The farther one looked, the greater was the speed of separation and, of course, this view delved farther into the past, since light had taken billions of years to travel from these far components of the Universe.

Other scenes appeared as the aloof Mind continued to muse on the content of the Universe. Groups of galaxies were pictured. Some of these took the form of spiral wheels like the home galaxy, some were large and apparently featureless elliptical galaxies. Some small fraction was quite irregular, showing signs of past or present collisions between whole galaxies. One pair in particular attracted Ali's attention. Long curved streams of stars sprayed out as two rotating galaxies interacted like colliding buzz saws.[15]

The vision presented to them, as they peered within the vastness of the Mind, flitted from one strange-looking feature to another. There were expanding cartwheels of matter, residues of vast supernova explosions. They saw distant galaxies, which had small star-like objects in their centers—quasi stars, or "quasars" that on their own pumped out more radiation than many complete galaxies.

They viewed one of these active galaxies more closely and saw that in its center there was a ravening core that drank down whole any star that chanced close by. From this core shot out two great lances of particles that extended for hundreds of light years on either side, racing away almost at the speed of light. It was a true Death Star.

The Mind's vision moved from the particular to the general and showed ever-greater regions of space. Individual galaxies, however peculiar, could no longer be resolved and attention was now on groups and superclusters of galaxies. These were concentrated in certain regions of space, along webs separated by immense voids. On this largest scale the visible matter in space

. . . .

15 This exists and indeed is the collision of two galaxies. The combined object is known as the "Antennae" because the long streams of stars look like two curved insect antennae.

was arranged as a sort of sponge that was pierced through with holes, or as a vast piece of lace whose every tiniest stitch was a galaxy containing hundreds of billions of stars.

Throughout this entire region flowed the Cosmic Background Radiation, the frozen relict of a great blast of light. Initially it was hotter and more brilliant than the surface of a star, but the expansion of the Universe had stretched its wavelength, lowered its frequency, and cooled its energy so that it was now unthinkably cold. It had initially filled all of space and so it still did, flowing equably in every direction as space expanded enormously around it.

Ali suddenly found this staggering vision was more than he could bear to contemplate and quietly backed away from the Mind's Eye. He left his companions still entranced as he crept out of the chamber and wandered aimlessly down featureless rock tunnels. Eventually, round a corner, he saw daylight pouring in from an opening to one side. After he passed through this he both heard and felt a deep rumble as plummeting rocks closed the opening behind him. He turned and found himself facing the same stony visage that had admitted him to the cave. With a noise of mountains splitting, this countenance raised one quizzical granite eyebrow.

"Well now," it grated, "have you enjoyed your visit? Have you discovered within treasures of revelation about the scale and nature of the Universe that you inhabit?"

"Yes, it was very interesting," admitted Ali, and then he was struck by a thought. "You wouldn't happen to have a cousin nearby who acts as a doorway to another cave would you?" he inquired hopefully. "A door that opens into a cave crammed with gold, perhaps?"

INTRODUCTION TO
the fourth tale

"So now I have told you something of the nature and scale of the Universe," remarked the Storyteller, carefully straightening his cravat. "I have described stars and galaxies, galactic clusters and the great filaments and voids that fill the Universe out to the very limit of vision.

"I admit that I may not have told you everything of how the Universe is *now*, but I have attempted to describe how you may *see* it now, which is a quite different thing. Light from its various distant parts finally arrives at your eyes only after long delays. The vast reaches of space are terrifyingly empty, occupied for as far as you may see by just a few blobs of matter–by stars and galaxies. However, as it is on one of these isolated blobs of matter that we live, they are of some interest to us.

"Matter makes up the stars and planets, together with any people that live on them. Matter, on the large scale, is held together by gravity, and gravity plays a vital role in controlling its distribution and its motion. It was gravity that made a newborn Universe of gas condense and curdle into *stars and galaxies. Gravity determines the shape of the whole Universe.*

"Gravity tells matter how to move, and it also tells space how to bend. Gravity makes SpaceTime curve in long billowing folds that hold the stars and planets."

"How can space bend?" asked Rachel. "There is nothing in empty space. You cannot bend nothing!"

"There may be no matter present, but that does not mean that space is nothing," replied the Storyteller. "Space and time form the framework of our experience. It is not space alone that we need consider. It is through the combined framework of SpaceTime that a thrown ball travels. If the framework of SpaceTime were straight then the ball would travel in a straight line, but the frame is bent, SpaceTime is curved, and your ball will curve round and fall back to Earth.

"That is what you see. A ball will fall back to Earth; a planet will orbit in a curved path around the Sun. They are all following the most direct path available to them in a curved SpaceTime. Looked at from the viewpoint of space alone it seems to you that a force is acting on them to deflect them from a straight path, and this force you know as gravity.

"Sit quietly and I shall tell you the story of 'Jack and the Starstalk.'"

the fourth tale:

JACK

and the

STARSTALK

(*spacetime* and *gravity*)

Jack walked along the country lane, leading Bossy the cow behind him. Leading is perhaps not the right word; more accurately, he meandered from hedgerow to hedgerow, pushing and pulling with all his might to try and coerce his companion away from particularly attractive clumps of grass and cow parsley. As he was occupied in applying his shoulder to the cow's angular and unmoving rear quarters, he heard himself accosted.

"Ah, good morrow, young sir! I perceive that you are having a little inconvenience with that no doubt otherwise excellent bovine quadruped."

Jack looked round and saw a tall flamboyant stranger who was partly obscuring a smaller associate lurking behind him. He was elegantly if a little shabbily dressed. He wore a top hat, going a little green in places, a frock coat in similar condition, and a flowing cloak with a few mismatched patches. Jack agreed that the cow was indeed a right pain.

"I have no doubt that you are conducting the animal to an appropriate place of commerce where you would hope to exchange her for specie, for coin of the realm. Might I be so bold as to put to you the proposition that, rather than endure the unpredictable rambling of your companion for the all-too-lengthy remaining distance to the neighborhood marketplace, you should allow us to relieve you of this irksome requirement."

"Pardon?" said Jack.

"I wish to buy the cow!" his new acquaintance replied more curtly.

Jack felt unsure. His widowed mother had told him specifically to take Bossy to the local market, but on the other hand it was still quite a long way, and he was becoming increasingly tired and frustrated. This friendly man used such long words that Jack felt he could surely trust him.

"How much money would you give me?" he asked shrewdly.

"Ah, money. The universal medium of exchange, society's vehicle of reimbursement for goods and service. You are possessed of some commodity; you exchange it for money and then consider yourself free to exchange this for any other article that you desire. But tell me," he asked earnestly, fixing the lad with a look of practiced sincerity, "have you considered cutting out the middleman? Certainly you may have money," he exclaimed dramatically if not entirely truthfully. "Money that you may be able to *exchange* for the article you desire, provided you can find someone who possesses what you want.

"Have you considered the comparative simplicity of exchanging whatever you wish to trade for the wherewithal to *construct* whatever you may wish? You have the option of possessing the actual components of any object whatsoever you might possibly desire. In exchange for your fine beast I am prepared to give you," he paused dramatically and held up a small bag, "this purse which contains a selection of quarks and electrons. From these one may construct *anything*."

Jack looked a bit dubious. "Are you sure that I could make anything I want from whatever is in that little bag? Are the quarks magic, then?"

"My dear boy, your doubts cut me to the quick. I can inform you without any fear of conceivable contradiction that any object you might desire will contain quarks. All quarks are pretty magic in their way, and if these should fail to provide whatever you may need and desire, then my name isn't Truthful John Certitude!"

"But your name *isn't* really Truthful John Ce... OUCH!" His smaller companion interrupted, but stopped abruptly as John made a sweeping gesture with his silver-headed cane, and the heavy if slightly tarnished knob accidentally impacted on his associate's temple.

Jack still felt a little unhappy about the offer. "That bag looks pretty small to me. I don't see how it can hold enough quarks and electrons to make anything very much at all."

"Such misgivings are, I suppose, to be expected from the unsophisticated," replied the would-be trader with a pained expression. "You must realize that quantity is a comparatively secondary property of quarks. Quarks, and indeed electrons, are absolutely identical. One electron is quite indistinguishable from another electron. One quark of a given type can in no way be identified as differing from another. There is no way that you could differentiate between the quarks that are in this bag and all the others in the world. Because of their total identity you cannot say that any of the other quarks in the world is not the same as any of the quarks in this bag. So you might, with all plausibility, say that you have all the quarks in the world in your bag. A number quite sufficient to fulfill your desires, I warrant."[1]

Jack was impressed. He didn't quite follow what Truthful John had said, but it did seem as if it should be convincing. He didn't quite follow the bit about number not being significant. He felt pretty sure that having one cow was not the same as having two cows, and felt a strong suspicion that having one cow was not at all

• • • • • • • • •

1 This is not a good argument. The number of quarks present is a perfectly meaningful quantity, but then you shouldn't believe what Truthful John tells you.

the same as having no cows, as was about to be the case. However, he didn't like to argue with such a plausible fellow, and it was after all a long way to the market, so he obediently handed over Bossy's halter in exchange for the purse of quarks.

He set off homeward, leaving Truthful John leaning on his cane and twirling his mustache in a satisfied way while his helper attempted to entice the cow out of the adjacent hedgerow. When Jack got home, he told his widowed mother what he had done, but she did not seem very impressed by John's promises. There was a period of lively discussion, during which the purse of quarks was thrown out of the window, and afterward both retired to bed in an unhappy mood.

Jack's mother looked out of her bedroom window as she wondered what she was to do with the lad. Dusk was gathering fast, but she could still make out the dark shapes of cows browsing on the meadow in front of her. It soothed her, in times of stress, to watch the herds that she had inherited from her late husband Wurzle MacDonald, the West of England cattle king. Oh well, it was obvious that Jack was not cut out for the marketing side of the business, but undoubtedly something could be found for him.

As night fell and Jack and his mother both slept, there was activity in the darkness outside. At first it was on far too small a scale to be perceived. Quarks and electrons assembled to form atoms. More and more and more came together until their conglomerate mass was substantial—and still they gathered. Were they the quarks and electrons from Jack's purse? Who can say? Quarks and electrons are indeed in no way distinguishable, and in this respect Truthful John would have been startled to discover that he had spoken no more than the truth. More and more and more, the bulk grew explosively until there was a mass of quarks and electrons that was sufficient to create a vast nugget of gold far greater than the total holdings of the Bank of England or even of the semi-legendary Fort Knox, in far-off America.

Unfortunately, they didn't do that.

Nature is, alas, a law unto herself and gold is, after all, a very simple substance. Nature likes a challenge and so the atoms formed from the quarks and electrons assembled themselves into the huge and intricate molecules of organic substances. When Jack threw open his curtains in the morning, he discovered not a gold mine but a huge twining trunk whose roots almost entirely enveloped their house. It rose above the rooftop and soared without hesitation into the overlying clouds. The very presence of this ladder to the stars was a clear invitation to Jack that he should immediately climb it. His mother, who had now awakened also, pleaded with him not to do anything silly. Then she realized that this was pointless advice to the hero of a tale like this, so she prepared for him a lunch of beef patties in a bun and waved him on his way as he began to climb.

Jack climbed steadily and the ground dropped away beneath him. He felt the satisfying solidity of the rough bulk under his feet and the twining shoots that he grabbed as he ascended. As he rose higher and higher, and particularly if he was so unwise as to look down, he became the more aware of gravity, and somehow the solidity of the trunk became less and less significant. He reached the level of the cloud cover and continued to climb doggedly through it. The damp clouds chilled him as he clambered determinedly upward but eventually he was above them and out in the sunlight. Still he climbed up this incredible plant that rose seemingly forever into the sky. Now the solidity of rock and earth was a thing of the past and the force of gravity increasingly dominated his thoughts. Upward, ever upward–would his climb ever come to an end? As with all things it did and he stepped from the topmost tendrils of this unbelievable plant to stand on–what? There should be no solid ground so high in the sky; he knew that. He had heard tell of "higher planes of existence," and perhaps he was standing on one of those. He looked around.

There was not a lot to see. The landscape (if it was land) was rather featureless. There were no hills, no trees, no tinkling little brooks. What there was nearby, and he could hardly avoid noticing it, was a great, massive castle, straight out of some Arthurian legend. Jack walked toward the tall entrance and across the heavy drawbridge, passing a sign that read "Giant's Castle." The heavy door was slightly ajar and to either half of it were pinned notices: "Quiet, I am thinking–Do Not Disturb!" and, more simply,

"SILENCE!" Jack had not, however, come so far to be put off by a
notice and went on in.

He found himself in a huge
banquet hall. There was a great
table in the center and seats for
many people around it. Or perhaps they were not for people as
they were far too large. The room
seemed to be empty, though as
the top of the table came up to the
level of Jack's eyes, he could not
clearly see what was at the far
end of it. He walked round the table and found at its head a massive carved chair, larger by far than the others. Its heavy back
ascended far above the level of Jack's head. At first glance he
thought this chair also was empty, but then he saw a chubby little
man perched in the middle of the great seat with his legs dangling
over the edge. He wore large, round spectacles and beamed at Jack
as he approached.

"Who are you?" asked Jack. He believed in the direct approach.
"You are not a giant!"

"Oh, but I am," replied the little man rather indignantly, drawing himself up to his full height so that he could almost see over
the arms of his chair. "I am an Intellectual Giant."

"Oh," said Jack. There didn't seem to be any good response to
that, so he changed the subject. "Don't you feel a bit nervous living
up here, so far above the Earth? Don't you worry about the force of
gravity?"

He had apparently chosen the right approach. The little man's
frown cleared and he positively beamed as he began to expound
his topic. "Ah, now then. When you say 'the force of gravity' I might
take exception to the implication. Gravity is not exactly a force in
the way that electrical attraction is a force." He squirmed into a
more comfortable position in the seat. "With an interaction like
electromagnetism you may get forces. Usually they depend on
some sort of charge, electrical in that case, so you have forces
acting between charged bodies. Depending on the charges on the
two bodies, they may either attract or repel one another.

"Now the thing about a force," he continued, getting into an
even more comfortable posture, "is that it may cause whatever it
acts on to move. If you push something, then it may well move, and

the heavier that something is the more reluctant it will be to move very quickly. If you push a child's toy cart, you can probably move it without trouble, but if you push a railroad car it will move very slowly if at all. The way that things will move depends on how heavy they are. Indeed, the very direction in which something moves will depend on whether the force acting is attractive or repulsive, and electrical forces may be either. You may, if you wish, consider gravity as a force," he granted generously, "a force that decreases in strength as the square of the distance separating interacting objects, just like the electrical force. It is a strange force, though. With the electrical force the strength will vary depending on the amount of charge that an object may happen to carry. You have opposite electric charges in matter, and they largely cancel one another out. Where the charges of one sign would repel some external charge, the charges of the opposite sign attract it, and the matter behaves as if it does not carry any electric charge at all.

"With gravity it is different, quite different. Different objects are all acted on by gravity. There is no cancellation among the interactions—they are always attractive. The objects concerned are all affected in *the same way*. It doesn't matter what they are, they will all move in the same direction and they all accelerate *at the same rate* whatever their mass may be. Sometimes it may seem that lighter things are less strongly influenced. A feather will fall through the air more slowly than a billiard ball, but this difference in behavior is not because of gravity. It is because of the air. When something falls through the air, then the resistance of the air will exert a force on it and this will slow a light object like a feather much more than it will a billiard ball. Gravity, you see, is not so much a force that may result in accelera-tion, as it is itself a pure acceleration; the same for every object concerned.

"Here help me down from this chair," he said abruptly as he paused in this lengthy exposition, "and I shall demon-strate what I mean." Now that Jack came to think about it he realized that his compan-ion could probably do with some help, as his feet did not come anywhere near the floor. With a little effort from the two of them, he managed a safe, if slightly ungainly, descent. Once he was upon the

floor, he led Jack from the hall and through a corridor to a courtyard outside. On the way he gestured at a number of tall figures lining the corridor, which Jack had taken initially to be decorative suits of armor, though they were remarkably featureless and not in the least decorative. Somewhat to his surprise each in turn stepped away from the wall as they passed by and all followed their master outside.

In the center of the yard was a smooth, square area. Very smooth it looked; Jack could not decide what sort of coating had been applied to it. His companion gestured in that direction. "There you have a frictionless surface. You do not come across these very often since there is always some atomic interaction between your feet and the ground you stand on that prevents them from slipping too readily. Indeed it would be impossible for you to walk if this were not so. Step onto this zone and see what I mean."

Jack, as may have been noted, was easily led and so did as he was instructed. He took one step onto the area and found to his surprise that he immediately lost all control over his movement. To say that it was slippery underfoot would be a gross understatement. His foot found no resistance at all. He could not push forward, turn aside or thrust down to alter his balance. He had stepped onto the slick material with a slight forward motion and this same velocity was preserved unchanged as he sped over the yard, despite all the waving of arms and frantic contortions that he attempted. He moved straight across the smooth area with no deviation or change in his speed until he was caught at the far side by one of the faceless attendants.

"That is how you may expect to move in the absence of applied forces," remarked his intellectual guide, walking carefully around the edge of the yard without standing on the smooth area. "As you no doubt observed, you moved across my yard in exactly the way that you started out. That was what you experienced on a flat surface without friction. Precisely the same would apply in a flat three-dimensional space. If there are no forces acting on you, you will move along in whatever direction you start off without any deviation or change in speed. You have to be careful if you jump away from your ship in empty space, as you would have no way to turn round and go back. That is what happens in a flat space," he emphasized.

"What is a flat space?" asked Jack, who thought that space was space and that was it.

"It is space that is not curved," answered the Intellectual Giant unhelpfully. "Step through here and I may be able to explain further," he continued, opening a door at the side of the courtyard and ushering Jack inside. Jack rather thought that he had not explained anything at all yet, but went through the door anyway. He found himself in a small, windowless room or closet. It was quite unfurnished, and indeed there was little room for any furniture. Two opposite walls were heavily padded with thick, soft ropes stretched across them, and there was a panel of buttons on the wall between. Otherwise the room was quite bare. His guide followed him in and as the door closed he pressed one of the buttons. The floor dropped away.

Jack whirled round in surprise and found himself floating in air in the middle of the room, moving slowly toward the far wall. "What has become of gravity, why aren't I falling?

"Ah, but you are—and so is the box you are in. You are in free fall under gravity, and so naturally you detect no sign of gravity. You are going freely where gravity leads you, as is everything in your vicinity. As everything falls in just the same way, it seems to you as if nothing falls at all because there is no relative movement for you to see." As his mentor was speaking Jack had reached the far padded wall, rebounded gently from it, and drifted back toward the one opposite.

"I would suggest that you hold on!" Jack's guide said suddenly, interrupting his own flow of information. Jack grabbed at the ropes and was abruptly slammed into the padded wall that had now become a floor, and he was flattened against it as if by a whole herd of his mother's cows sitting on top of him. The period of intolerable weight was brief and then he found himself floating as before.

"As you could feel, we have just changed our direction of motion. You are again in free fall, though to an onlooker it might not seem so. Observe!" He pressed another button on his control panel and a shutter slid aside in what had initially been the floor but was now just another side of a box that had neither up nor down. The shutter covered a window and through the window Jack could see the surface of the Earth. Ocean, coastline, tiny towns and fields, all sped past the window in a continuous remote stream. "We are still falling, but not straight toward the Earth as we had been. We are now falling sideways, or rather we are moving sideways while steadily falling. We are in an orbit round the Earth. Our speed would tend to carry us away from the Earth entirely, but our continuous fall bends our path down toward the surface so that we move in a circular orbit, not flying away but neither coming any closer. We are in a very low, fast orbit that will soon bring us back to our starting point. Normally it would not be practicable to orbit so close to the surface because air resistance would slow us down, but as a theoretician I am accomplished at ignoring air resistance.

"Brace yourself!" he commanded suddenly. Once again Jack slammed into a padded wall, though this time it was the opposite one. The force continued for a time and then he was standing on the floor, now covered again with its shutter. "Now we are not falling. The force you feel on your feet comes from the atoms in the floor as they interact with atoms in your shoes. This prevents you from following the path that gravity would dictate, but the force you feel is from those atoms and not gravity."

The door slid open and they stepped out into a vast and formless room. I misinformed you slightly," admitted his guide. "It is not space alone that is flat or curved. It is SpaceTime." Before Jack could ask the obvious question, he went on. "I realize that I need to say something about SpaceTime, be it flat or curved, so I have brought you to a place where we may try to visualize it. Behold the light cone!"[2]

The whole expanse of space before them seemed to shrink down into a flat plane and above this, in some new direction that was not quite space, soared a great cone that was matched by another complementary cone stretching down below the plane. The cones had as their axis a thin line that ran straight up out of sight.

"That is the direction of time. It is in some ways a bit like yet another direction in space, though it is certainly not the *same* as space.[3]

* * * * * * * *

2 The light cone in flat SpaceTime has already been discussed in the first tale, "The Prince and p."
3 In its mathematical representation the time axis is said to be imaginary. It is imaginary only in the technical mathematical sense of including the bizarre factor "i" (the square root of –1). Time is, of course, all too real.

• •

The Light Cone

The light cone was introduced in the tale of "The Prince and p," but it is worth reminding you of it. It attempts to portray the three dimensions of space and also the extra dimension of time. It is impossible to draw a diagram in four dimensions, so only two of the three dimensions of space are included.

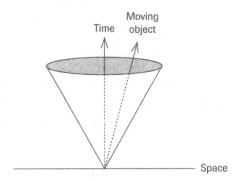

The light cone is really an expanding sphere in space, showing how light moves out from the origin with increasing time. All you can reasonably draw is a circle that spreads as it moves onward in time and so traces out a cone. Even the cone must be drawn in section when it is shown on a flat sheet of paper.

A path inclined to the time axis represents the time axis of a moving object. Its time axis is slightly space-like; in the sense that at different times it is in different positions. That is a fairly basic description of movement.

The path for anything that moves at the speed of light would lie along the light cone. Nothing can go faster than this.

• •

"Matter, the things that make up our Universe, must be described both in space and in time. Objects not only exist, but they move about and the intricacy of their movements contributes to the richness of what we see about us. For that matter, even if an object were to stay still, it would still need time in which to exist. The movement of objects is constrained and limited, and the light cone shows the limit. There is a maximum speed with which anything may move. It is usually called the speed of light, though this special speed does not have anything in particular to do with light.

It is a property of SpaceTime itself and it just happens that light moves as fast as it is possible for anything to move.

"In the representation of SpaceTime that is displayed here, slanting lines show objects in motion. Since the depiction shows the positions of things at different times, it follows that a slanting line represents something that changes position with time, something that is *moving*. A small slope from the vertical belongs to something that is moving but slowly. The greater the slope, the faster the speed until you reach the maximum possible and the path lies along the light cone, which contains all directions of movement at the speed of light. Nothing can move faster than this, nothing can move outside the light cone.

"This diagram shows *events*, happenings that each have a definite location both in space and time. It is SpaceTime from one point of view. That point of view is *yours.*" The shining cone moved abruptly to focus itself upon Jack. He found that now the cones met where he was standing and the time axis stretched away ahead of *him.*

"You have your own light cone, the only one that really concerns you. It is centered on your own personal 'here and now.' Within the upper light cone lie all the events that you might experience in your future. On the light cone lie all the locations that may see an image of you as you are at this very moment, intercepting the light from you as it spreads out through space in a great expanding sphere. In the lower light cone lie the possible events of your past, and the lower cone has on its surface all events from which light may reach you *now.*"

One of the Intellectual Giant's faceless attendants came to stand silently a little way off. From and around its location another double light cone now extended. "An observer standing off to one side of you will have his or her own light cone and it will be slightly different from yours, since your position will not be at the center of things for that observer."

Other metallic figures strode out and took positions so that they formed a line, each with his own SpaceTime represented by light cones projecting into his personal future and past. The cones from the more remote figures intersected Jack's time axis farther up, to show that light from them would take that much longer to reach him.

"If there is no relative motion between any of you, your axes of time will all be in the same direction and any moving object will be found by all of you to have the same speed."

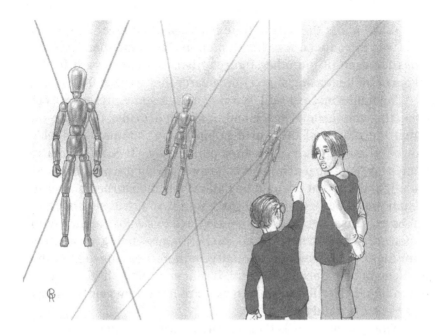

"So I should think!" said Jack roundly. "Time is time after all. It is obviously the same for everyone." Jack might be easily influenced, but he still felt he knew his own mind on such things.

"That turns out, in fact, not to be the case," replied his brainy escort. "A moving observer will find velocity and even time to differ from your view of it.[4] SpaceTime differs for different observers. What is of more relevance to us is that SpaceTime will also be affected by the presence of mass. SpaceTime and matter are together embedded in the Universe and each affects the other. A mass such as a sun or even a planet will pull on the web of space and time and cause them to distort so that they become curved. Let us study what happens when there is a mass nearby."

As if the Intellectual Giant's merest whim was law, which in this hall of demonstration it was, a great shape appeared in the distance. This represented the presence of a planet such as the Earth but was not spherical in appearance. Since three-dimensional space had been reduced to but two dimensions so that time also might be portrayed, the planet was a circle in space. It showed itself in SpaceTime as a tall cylinder since its shape was

• • • • • • • •

4 The distortions of space and time when seen from moving frames are spoken of in the first tale, "The Prince and p."

unchanged along the axis of time from past to future. When this mass was introduced into the depiction, Jack saw that the light cones for each of the other figures tilted slightly toward it, the degree of tilt increasing slightly the nearer they were to the planet.

"The light cone, as I said, is a way of trying to represent Space-Time, the composite of space and time that provides the stage on which the drama of the world is played out," remarked Jack's avowedly intelligent guide rather pompously. "It is not a totally empty stage and SpaceTime plays its own part in the performance. You might think that it would be natural for undisturbed matter to be stationary, but when you think about it you, will see that this makes no sense, because you would have to say 'Stationary with respect to what?' There is no such thing as being absolutely motionless. All movement is relative and nothing may claim the privilege of being absolutely still. If matter is left to itself, then in all probability it will be seen to move. From the viewpoint of many observers it certainly will move.

"What you may say is that it is the nature of matter to move in as straight a line as and at as constant a speed it can. When you moved across the frictionless center of my courtyard, you moved in a straight line at constant speed and if there had been no farther edge, you would have gone on indefinitely, just as long as we could neglect all forms of friction. And I am well practiced at neglecting friction," he added as an aside.

"If that frictionless surface had been on the surface of the Earth you would have sped along with a constant speed, but it would no longer have been in an absolutely straight line. The Earth is a sphere and so you would have to follow a curved path because the surface on which you travel is curved. The straightest line that you can trace around the Earth is called a 'great circle' and if you could truly ignore all forms of resistance, this 'straight line' might eventually bring you back to where you started." The self proclaimed intellectual giant paused for a moment with his usual air of satisfaction and then continued.

"Perhaps you are thinking that you can readily see that the Earth is curved but are still unconvinced that empty space might be curved also. After all, there is nothing there to curve is there? I have several things to say in answer to that." Jack had rather feared that he would have. "It has not always been obvious that the surface of the Earth is rounded. Though this was clearly shown in ancient Greece, it was not general knowledge until much later and there are still people who would dispute it. On the other hand it is not space alone that is curved, it is SpaceTime: the composite of space and time that controls how matter may move and develop. SpaceTime is not an inert stage within which the action develops quite independently; in fact, SpaceTime tells matter how to move. The relationship is reciprocal as matter tells SpaceTime how to curve. It is the presence of mass that distorts SpaceTime and the greater the mass the greater the distortion.[5] It is this curvature that you call gravity," he finished abruptly.

● ●

Gravity and Curved SpaceTime

Newton saw space as *flat*, the space of Euclid. It is not easy to visualize odd geometries in three dimensions, let alone in the four dimensions of SpaceTime. In two dimensions a Euclidian plane is like a flat sheet of paper. It is fairly easy to visualize a curved two-dimensional surface, like the surface of the Earth. If you were to travel over the Earth in as straight a line as possible you would eventually come back to the same point, having traveled right around the Earth.

Newton stated that in the absence of any forces, an object would travel in a straight line at constant speed. Something like a planet is traveling in as straight a line as it may, but in a curved SpaceTime this means that it travels in an orbit around the Sun. Newton's view was that mass such as the Sun would generate a *force of gravity* that bends the path of a planet into an orbit around it. In Einstein's view the presence of mass distorts the structure of SpaceTime and a planet follows the straightest possible path through this curved SpaceTime.

The diagram illustrates how the distortion of SpaceTime around a planet will affect the motion of a falling object. The object would see itself as stationary and moving into the future without changing position. (You do not ever see yourself getting nearer or farther away from yourself, though other object's, like the surface of the Earth, might be rushing toward you.) An outside observer, one on the planet, say, would however see the object's SpaceTime as tilted and so it would be seen as changing position with time. It would be seen as *moving*.

● ● ● ● ● ● ● ● ●

5 Why is this so? Good question. I don't know. It seems to be how things are.

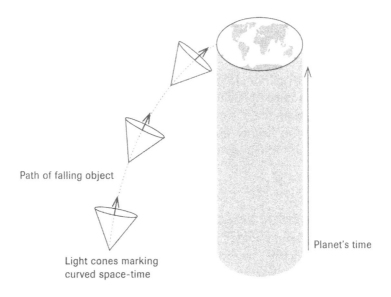

Path of falling object

Planet's time

Light cones marking
curved space-time

(Angles increased about a million times)

In the curved SpaceTime around a planet the tilt and thus the observed speed will increase as the object moves toward the planet. This is seen as the acceleration of gravity. The distortion of SpaceTime around the Earth is very small and the tilt of the light cones would be very small. The vertical scale in the diagram should be expanded about a million times.

•   •

"But how can that be?" protested Jack. "Gravity makes you move faster and faster as you fall. Where is the curvature of space in that?"

"I shall try to show you. You see before you a series of token observers, represented by my faithful attendants. Each in turn is closer to a planetary mass than the one before and so you see his light cone tilt a little more toward that mass. This demonstrates how space is curved. If you had a line of ships well spaced out across a major (but remarkably calm) ocean, the mast of each would be tilted just a little with respect to its neighbors on either side as they range around the curvature of the Earth. In that case their masts would point toward the sky. In the case of the light cone it points toward the future, but in each case the tilt indicates curvature. Note that is how *you* observe it. For each observer his own light cone is upright, of course, and the axis of time points resolutely into the future as it would in a flat space. As far as he is concerned it is everyone else whose light cones are tilted relative to him. He himself does not feel gravity. That is for other people to notice and you, as an outside observer, are one of those people.

"When you observe someone who is closer to a large mass than you are, you will see his light cone as tilted over. From his point of view it will not be, but for you it will. The thing about relativity is that it is relative. Observe what happens as time passes, and remember that it is in *your* time that you observe, not in his. In his own time he is standing motionless but from your perspective his time axis is inclined to yours[6] and so you see him as moving steadily away from the motionless state and as moving ever faster. You might well describe what you see by saying that a gravitational force is accelerating him.

"As your time passes, you see his speed to be increasing. He will see things differently. In his own time and place he will be always at rest. Come and see for yourself. We had better move smartly though, since he is getting away." The Intellectual Giant led the way over to the nearest of the statue-like observers. He had previously been receding from them at an ever-quickening pace, but they managed to overtake him. Soon enough they arrived at his position and Jack perceived that the axis of time was now soaring straight up into the local future while the light cone that had previously been his now appeared slightly tilted.

"As you see, any observer will perceive himself or herself to be at rest. Or in this case, itself," he added as an afterthought, indicating the faceless metal observer beside them. "Wherever you are, you see yourself in calm, normal surroundings with no sign of any 'force of gravity.' As you look around you may see that others are accelerated relative to you, but that is someone else's problem, is it not? When you stand upon the surface of the Earth, you might think that you feel the force of gravity, but what you really feel is the force exerted by the Earth itself. You feel the Earth pushing up on your feet. The atomic forces that make matter behave as solid resist your free motion. They prevent you from moving along the straightest path through a curved SpaceTime. If you are in 'free fall,' then in your own time and space you are still, and it is everyone else who is moving. See how it would look in your normal view of space and time!"

· · · · · · · · ·

6 I do hope that you are remembering to think in four dimensions throughout this.

The vast simulation of SpaceTime faded from Jack's sight and he found himself looking at an even vaster scene. Black empty space extended all around him, save in one direction where he saw the great and steadily expanding globe of the Earth. He and his two companions, the metal observer and its small rotund master, were floating calmly together with no impression of movement, but they could see, filling almost half their field of view and rushing headlong toward them, the cloud covered surface of a continent below.

"As I said, in your own proper frame *you* are not moving, but other things may be. It might be best not to hang around until the Earth's surface arrives here. You may not feel any force of gravity, but the inter-atomic forces between the ground and yourself could be, shall we say, unpleasant. Come away!"

Jack felt himself plucked away from the path of the onrushing planet and found that they were again standing in the lofty hall. His host continued his discourse. "That is gravity. It is a distortion of the fabric of the space and time within which objects must move, and so it controls their movements. Where you have mass you have such distortions. Normally these are tiny. Even around something as seemingly massive as the Earth that you walk upon, that seems to hold you to itself with such a firm grip, the tilting of the light cone would appear infinitesimal. It would appear less than the breadth of a hair as seen at arms length. The effect of the Sun's mass is not so very much greater than that of the Earth, even though the Sun is significantly bigger. But the minute distortion of space that such mass produces is sufficient to guide the planets in their orbits around the Sun and to govern the fiery fall of a meteorite, a so-called shooting star. Of course, since SpaceTime is but slightly distorted, the velocities involved are very low."

"How can you say that?" objected Jack. "When I see a shooting star streak across the sky it is clearly moving quickly!"

"That is relative. Whether a speed may be considered large or small is in general relative, but in the case of relativity there is a speed that is not in fact relative. The speed of light is absolute. It is the limiting velocity for any object that moves within SpaceTime. When we compare the directions of space and of time, as I do when I speak of the tilting of the time axis, then we must consider that a second of time is equivalent to a light second of space, and that is roughly the distance from the Earth to the Moon. On such a scale the velocity with which a body may fall to Earth would show in SpaceTime as a line scarcely separated from the time axis. The velocity that you mentioned before falls far, far short of what would be needed to take you from the Earth to the Moon in a second." Jack looked again at the series of light cones represented before him and indeed it now appeared that the relative tilt of each one was so slight as scarcely to be detectable.

"In such a routine case the view of gravity as a distortion of space and time gives very much the same results as does Sir Isaac Newton's grand notion of the *force* of gravity. However, although Newton's theory of gravitation may appear fully adequate within the Solar System, it is not so everywhere. Sometimes in the Universe you *do* meet with a strong curvature of space and time. You have Strong Gravity."

At this dramatic utterance the light cones associated with the token observers all tilted further from the vertical. Now the angles of tilt were appreciable and growing steadily as Jack looked from one to the next. For the most remote observer his cone had tilted so far that even the *nearest* side of his light cone was tilted away from Jack. If Jack understood what he had been told about this display, it suggested that not only would gravity be extremely strong there but also that no light should be able to reach Jack from such a position.

The plump little Intellectual Giant regarded his display with some satisfaction. "It is the presence of mass that makes space curve, it is mass that creates gravity and so pulls other mass toward it. In order to produce such a great distortion of space and time we need a lot of mass, but it is not just the amount of mass that is important. How it is distributed is also important. According to Newton, gravity is an *inverse square force*. If you have a large, round, heavy object, like a planet, then the gravity this produces will increase the closer you get and will increase accordingly as the *square* of the distance to the center *decreases*. If you are half as far away as before the force will be four times as much, a third of the distance and you have nine times the force. Of course, this only holds if you remain outside the planet. If you were to sink down a deep, deep mineshaft then the strength of gravity would begin to decease, because part of the Earth would now be attracting you from above. You can get a much stronger effect of gravity if the massive object producing it is small, because you can then get that much closer to it all while still remaining outside." The little figure looked somehow shrunken as he considered such crushing forces.

"It will be clear to you that the gravity produced by a planet will act not only on other planets and satellites, it will act also on the material of the planet itself. The inner layers will try to pull the outer layers toward the center. The effect produced depends on the amount of mass concerned. Come with me."

He led Jack across the great display hall and into another room that looked like some sort of small gymnasium hall. In the centre stood a squat figure with wide shoulders and heavy muscular arms. Around his rather sloping forehead was bound a cloth that carried the symbol 'G.'

"This is the Mass Crusher. He exemplifies the effect of gravity on a massive object." He tossed a ball of some sort to the Crusher, who caught it between his powerful hands and began to squeeze it. After some moments it became obvious that his efforts were having little effect.

"You see that this solid object resists the efforts of the Crusher, as you might well have expected. It is not even necessary that the object be solid, a liquid will resist compression almost as well and even a gas can be compressed only so far before it strenuously resist any further concentration."

"So it is the forces within the material that prevent it from collapsing," remarked Jack in an attempt to summarize.

"Well no, not exactly. The electrical forces between the atoms are on the whole attractive and would tend to help the concentration of the material. It is the Exclusion Principle that keeps the atoms apart.[7] If the atoms were superimposed, their electrons would be in the same state and that is not allowed. Quite large solid objects, like the Earth, are able to resist the crushing forces of gravity, but at the center of the Earth the rocks are crushed to a dramatically high density by the overlying mass. It is also rather hot.

"When gravity compresses an object, such as a large ball of gas, this heats up since the falling atoms are losing gravitational potential energy that converts to heat. It is, however, a rule with gravity that, given enough mass, gravity always wins, but not necessarily right away." The Crusher had grown to a great hulking figure and was bearing down heavily on a fuzzy ball of gas clenched between his forceful hands. Suddenly the object began to

· · · · · · · · ·

7 See the tale of Snow White. The exclusion principle prohibits two electrons from being in the same state and in effect this inhibits their states in different atoms from being in the same position. You cannot easily push one atom into another.

glow mightily and his hands were
forced back by the blaze of energy from
between them.

"A star, such as the Sun, is a massive object and it would be certain to
collapse under gravity save that this
very collapse produces heat that in turn
gives off radiation that will support the
upper layers. Initially the heat came
from the energy released by collapse
under gravity, but as the temperature
rose, it ignited nuclear processes and at
present the collapse of the Sun is held at

bay by the radiation from a raging nuclear furnace in its interior.

"This cannot go on forever. Eventually the fuel supply of even
the largest star must give out and it will again collapse." The light
between the Crusher's hands died and, with a grunt of satisfaction,
he brought them together. Whatever was now between them was
obviously minute.

"That is the typical fate of a moderate star such as your Sun.
When nuclear reactions die out, the radiation that is holding up
the gas envelope will fail and the star will collapse in on itself. For
smaller stars this collapse may eventually halt to give a white
dwarf. Things are different if you have a *really* large star."

The Crusher grew again to gigantic size. He had a glowing star
ball within his grasp but abruptly its light failed. He triumphantly
crushed it down to a tiny residue while the brilliant glare from a
supernova escaped from his clamped hands.

"When such a giant star collapses, the density of the mass
grows to such a value that nothing can resist gravity. Paradoxically
no repulsive force may resist because collapse against any such
repulsion will require energy. Now energy is mass and gravity acts
on mass, so once gravity is strong enough, any energy from the
repulsion that attempts to halt the collapse will in fact hasten it."

"So what does stop it from collapsing forever?" asked Jack.

"Nothing, there is no limit. The mass will shrink without limit.
It will become a *black hole.*"

"What is that?" asked Jack.

"A black hole is nothing, in a sense. Certainly it is nothing that
you can see. It is the monument that marks where matter has
shrunk without limit and left behind only a region of violently distorted SpaceTime that quarantines whatever is inside the *event*

horizon. No light may escape; nothing can escape from within a black hole. All sign of whatever it was that went into the black hole is gone and outwardly it is almost featureless."

· ·

Black Holes
(*The sink of all things*)

Gravity is a weak force between two particles and still fairly weak between planets. When there is enough mass and it is sufficiently concentrated, however, gravity becomes strong. Very strong. Eventually it reaches the point where no repulsive force can resist it. Indeed any resistance would only increase its effect, since resistance against collapse will build up energy, energy is mass and the extra mass would just increase the effect of gravity.

Around a black hole there is a border known as the event horizon. From within this border nothing can escape to the outside Universe, not even light. Hence the name black hole.

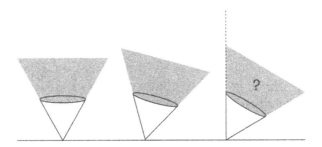

The SpaceTime diagram for an object falling into a black hole looks superficially similar to the one given previously for a planet, but here the angles have not been exaggerated. At the event horizon of the black hole the light cone is tilted so far that no light may escape outward (to the left).

I have not shown the time axes because I do not know how to draw them for this degree of SpaceTime distortion. Both the space and the time axes have moved in toward the light cone and at the event horizon an outside observer would see time stand still.

· ·

The Crusher opened his hands with an expression of surprise as he found that there appeared to be nothing left in them.

Between his palms was a tiny *something*: a warped region of SpaceTime that drifted out into the room. It collided with a set of weights and Jack was startled by the sight of heavy metal disks being sucked one by one into an infinitesimal point in space.

"Careful!" cried the Intellectual Giant. "That thing is dangerous! It might suck us all into itself." One of his selfless metal attendants sprang forward with a light cone held in front of him. He carefully aligned it so that the black hole lay just within the surface of the cone. As the black hole was now close to their light cone, this meant of course that it was travelling relative to them at almost the speed of light. It shot swiftly away from them. Soon it had vanished safely toward the depths of interstellar space.

When Jack looked around he saw that the Crusher had vanished. "You don't think...?" he said anxiously to the self-satisfied genius by his side. "No, I am sure he is quite all right. As he is a personification of gravitational collapse, it is not to be expected that he could be absorbed by a black hole. He probably had a pressing engagement somewhere else.

"Now let us for a moment consider further the effects of black holes—the conclusive product of gravitational compression, the ultimate in distorted SpaceTime."

Jack's diminutive instructor led him out of the great hall. Together they stepped through a door and, after a moment of disorientation, found themselves in the control room of a star ship. Anyone who followed science fiction would have recognized it as such. The captain and other members of the crew lay in acceleration couches, designed to cushion frail flesh and blood from any sudden acceleration. The one exception was the tall figure of the Astrogator, who stood immobile to one side. The vague metallic sheen of this figure suggested that he had no need to worry about the perils of flesh and blood. Everyone was facing a large view screen that was almost entirely empty apart from a dense field of stars. The center of the view was even emptier, being a blank black region where not even stars showed, and ringed with a number of bright streaks among the background stars.

"How did we get here?" asked Jack. "Surely we must be far into space and yet it seemed to take no time. Doesn't that mean we must have travelled faster than the speed of light?"

"It is not the speed of light that applies, but the speed of imagination. You do not imagine that this spaceship is real, do you? But it provides a useful way of looking at a black hole."

The captain spoke from his couch. "What is that in the forward viewer?"

"It is mass, Captain, but you cannot perceive it. It is a black hole that has taken mass from our Universe and hidden it away where we cannot see it." The tall rigid-faced Astrogator turned his head to look toward the Captain. This was a rather strange sight, as his body did not move at all. His head turned as if mounted on a pivot. He adjusted a couple of controls on a little panel in front of him and the empty center of the view was outlined with a dotted circle. "The boundary of the hole is the event horizon I have shown you here. It is the last point from which light may escape to the outside space; anything within is cut off from the rest of the Universe. You can no more examine the singularity within than you can discover what type of material initially produced the hole." The speakers perfectly formed mouth did not actually move during these remarks.

"What is a singularity?" Jack whispered to his guide.

"Just wait, the Astrogator will explain."

Indeed he did, continuing with as much reaction as if he were a pre-recorded message, "On a sufficiently grand scale there is nothing that can inhibit gravity. The material that went into producing the black hole has continued to shrink down, eventually to become a *singularity*: a point of no size and infinite density. Such an extreme object seems illogical." You got the impression that he would have raised one eyebrow in disdain, save that his face was totally immobile. " It is not something the Universe can be expected to deal with, but then it doesn't have to. It is hid from your view by *cosmic censorship*, safely within its event horizon, so that the cosmos may continue to appear logical and reasonable."

"Very well," replied the Captain, somewhat impatient at this lengthy reply since he was a man of action. " Put us in an orbit around that event horizon," he commanded the out-of-scene Engineering Officer.

"I cannae do that Captain. We do not hae the power to go anywhere near yon beastie and escape," replied the Engineering Officer over the intercom. It was clear that, in the time-honored tradition of ships' engineers, he was of the Scottish persuasion.

"Why, do we have a power failure?"

"No, Captain, there isna enough power in the Galaxy to do that."

"The problem Captain," interposed the Astogator efficiently, "is that, although the event horizon is the closest point from which light may escape, it may escape only in one direction: straight out.

Even light, if it were emitted along any other path, would be deflected so strongly that it would fall into the hole and never emerge."

There was a faint whirring noise as his arm moved across the control board. Now the display showed wavy lines leaving the event horizon, which was indicated by the dotted circle. One that headed straight out from the circle continued into the vast beyond, but at all other angles the light beams curved around and went back inside the hole.

"It is this effect on the paths of light that makes the hole black because light cannot escape from it. Equally it follows that beams of light that come from outside and are merely passing near it will curve to a lesser degree. This gives the effect of *gravitational lensing* that allows you in a sense to see something that cannot itself be seen. A black hole may be detected from the way that it can give multiple images of distant stars or galaxies behind it. You can see some of these images on the viewer.

"In principle it is possible to escape even from the event horizon of a black hole if you travel straight out, but you have to do so at the speed of light. Any slower and you are lost forever. There is no question of *orbiting* at that distance. The closest possible distance for a stable orbit is at twice the radius of the event horizon, and even there the orbital velocity is the speed of light, so any practical orbit must be further out again." A command from the captain caused the great spaceship to boldly, if rather cautiously, go and take up an orbit at a respectful distance from the event horizon of the black hole. The Engineering Officer flatly (if almost unintelligibly) refused to go any closer, so it was decided instead to launch a probe, since it would not be such a disaster if they could not recover it. They watched on the view screen as the game little craft plunged in toward the waiting black hole. It raced downward at first, and then as it approached closer to the event horizon, it seemed to slow. Its passage faltered and its image became redder. It no longer appeared to move inward but instead became redder and redder and faded from sight.

"What has happened to it?" asked the Captain. Jack had been about to ask the same.

"It is the effect of time dilation," replied the Astrogator unemotionally. "As the probe moves deeper into the distorted SpaceTime of the gravitational field, its time moves more slowly as we see it. We see that its motion slows; the frequency of the light from it decreases also and becomes steadily redder, soon passing outside of the range that we or our instruments may detect. It vanishes from sight, though in our reference frame it is still there. It would not be the same if we were riding down with the probe, as you initially intended Captain. We would then see the journey as one that continued down without pause, passing the event horizon without remark and quickly approaching the singularity. We would probably discover what happens at the singularity. I doubt that we would enjoy or survive the experience, but for sure we would never be able to tell anyone outside. It is a one-way journey."

As the crew stared regretfully at the point where they had seen the last visible sign of their probe, Jack's host whispered to him, "We might as well leave now." He led him away from the bridge and in a moment they were back in the castle, coming out of the same door that had led to the star ship.

"That episode was of course quite imaginary, a way of visualizing the presence of a black hole. To look at black holes in reality we must use light, or at any rate radiation of some frequency. To see the ultimate in gravity and curved SpaceTime you must look far out into space. Come, there is someone I should like you to meet."

Jack's scholarly host led him down a shadowy hall and ushered him through a door that led to a tight spiraling staircase. They began to climb and for a long period thereafter they continued to ascend–and continued and continued and continued...

Eventually they reached a tiny turret with a thick transparent window filling one wall. A great telescope floated in space, apparently just outside the window, pivoting on wide sails of photoelectric cells. Floating by it was a grotesque figure wearing some sort of helmet with enormous staring eyes behind great goggles. He had two gas cylinders strapped to his back from which flexible pipes looped around to his helmet. "Jack," his host said to him, "I should like you to meet the Hubblegoblin. He is the custodian of this Space Telescope, and with his assistance you can look far into the depths of space and see remote sights from the distant past."

"Pleased to meet you," remarked a speaker set in the wall of the turret. Despite what appeared to be a remarkable thick window, and one that apparently had no air beyond it, the voice came through perfectly clearly.

"Can I reply to him?" asked Jack.

"Yes of course, the communication works both ways," answered the Intellectual Giant. "All the information from the telescope is sent across to us by radio signal while our requests to direct the telescope in one direction or another are sent back by radio also."

"Isn't it a bit inconvenient to have your telescope out of reach like that? Wouldn't it be better to have it on the ground where you could work on it?" asked Jack.

"Oh no, not at all. It might be more convenient, certainly, but it would not be better. If it were on the ground, then it would be underneath the Earth's atmosphere and this is so turbulent, with winds and hot and cold regions and all that, that it distorts the light passing through. It is the atmosphere that makes the stars twinkle, but we want them to be steady so that we can see them clearly. In space there is less matter in the path of the light coming from distant galaxies, so we get less distortion of the image. That is not to say there is no distortion. The light has come a long way and it has passed through interstellar gas on its journey, but the image is still much better than you can see from below the Earth's thick blanket of air. What this interstellar gas does to the light as it makes its journey to us can be interesting in itself because it tells us something about the gas."

"It is not only gas that the light must pass through," interrupted the voice from the speaker, which had been silent for some time. "Some of the distortions that you may see in the images from my telescope are themselves relevant to your attempt to detect black holes. Observe this."

They could see the great telescope begin to swing around and then the outside view abruptly vanished from the window, which was thus revealed to be a large viewing screen. On it there was now a picture of a starry sky, the view as seen by the telescope. It grew in detail and sharpness as the instrument gathered more light from the distant scene it was observing. As always in such a scene there were many stars of wildly varying brightness, together with the fuzzy shapes of distant galaxies. In a rough circle around the center lay a sequence of curved bright streaks.

"There you have an instance of gravitational focussing of light," continued the Hubblegoblin. " What you see are images of a distant galaxy that is lying somewhere on the far side of a huge mass. As the light from the galaxy passes the mass on several sides it is bent so that it ends up in the telescope and you see many, if slightly distorted, images of the same galactic source. To achieve this effect you do need a lot of mass in the path of the light and the most plausible candidate for this concentrated mass is a black hole, and a mighty large one at that. We are not talking of the sort of black hole that is left after the death of a brilliant star,[8] but of one that is truly enormous, containing the mass of many millions of stars. Such very massive holes may be formed in the centers of galaxies, where they have absorbed much of its gas and even whole stars. They show dramatically the effects of a space strongly curved by gravity.

"Slightly less extreme objects than black holes may also show us the presence of novel, non-Newtonian effects of gravity. Observe, for example, a binary pulsar, a flashing star in orbit around a more normal companion. Many stars are held together in pairs by their mutual gravitational attraction. In this case one of the odd couple is unusual." The scene in the viewing window changed to show–well not a lot of anything really. There was a faint star in the middle of the field of view. "There you see a rare object." It did not look very striking to Jack, but he presumed there was something strange about it. "It may not look like much when viewed by the light that is visible to your eye, but observe how it emits when you observe it at microwave radio frequency."

As the commentary from the speaker continued, an area outlined by a rectangular frame appeared at the corner of the window, and within this could be seen a bright spot that flashed on and off with great rapidity. Some explanatory figures ran down the side of the area to specify the frequencies of the radiation and of the flashing.

"What you see there is a *pulsar*, a star that sends out radiation in short repetitive pulses. It is in fact a *neutron star*, a condensed star that has been compressed so much that protons and electrons have combined to form neutrons. The residual star is supported against further collapse by degeneracy pressure, much like the electron degeneracy that supports a White Dwarf star.[9] A neutron star has collapsed so far that it is tiny, only a few kilometers across, but still weighs rather more than a star like the Sun.

• • • • • • • • •

8 See "Cinderenda and the Death of Stars" later in the book.
9 Neutron stars are treated further in the later tale, "Cinderenda and the Death of Stars."

"In general such stars spin rapidly on their axis. The angular momentum that they had when they were normal full-sized stars remained when they shrank down to such a tiny remnant. Stars usually rotate fairly slowly, if only because centrifugal forces would strip off their outer layers if they spun too quickly, much as a loose fitting hat might fly off on a fast spinning carousel. Because angular momentum is given by the frequency of rotation multiplied by the size of the rotating object,[10] this tiny leftover object will be forced to rotate very rapidly indeed. It may spin round many times in a second.

"As the star collapses to a degenerate cinder and becomes a neutron star, its magnetic field will concentrate as it shrinks, and this intense field may produce a finely directed jet of radio emission. As the star whirls around, this beam of radio waves sweeps across the heavens like a frenetic lighthouse. For some neutron stars this rotating beam will sweep once across the Earth with every revolution and so appear to us as a sequence of short pulses.

"This whirling dervish of a star has a larger companion, as I explained to you originally. The neutron star is in close orbit around a more normal star so that the two form a close binary system, a very close system. The neutron star rushes round so rapidly in its tight orbit that the frequency of the pulses from its rotating beam, appear to vary as it goes around, simply because of the Doppler shift.[11] As the neutron star moves toward us in its orbit the frequency of the pulses increases with a *blue shift*. As the neutron star passes around its companion and moves away from us, the frequency is seen to drop with a typical *red shift*. This is because the neutron star is moving very rapidly in a tight orbit. This frequency shift tells us the neutron star is in orbit even though our telescope cannot separate it from the other star."

Another frame appeared in the corner of the window, this one showing a diagram of the tiny neutron star circling around its more normal companion. Below this was a trace that showed the series of bursts coming from the pulsar. As the source moved around in its orbit, the frequency of the pulses changed visibly,

• • • • • • • • •

10 The Dominie in "The Prince and ρ" introduces angular momentum.
11 This is both the frequency of the pulses and also the radio frequency. As the star moves toward us, the radiation it sends is compressed into a decreasing distance. This increases the frequency of *any* variation, whether the frequency of the radio waves or the frequency of the pulses of radiation. It is all speeded up in the same way.

being slower as it moved away and speeding up as the source moved back.

"So what has this to do with General Relativity, which is just another name for the theory of curved SpaceTime?" the Hubblegoblin asked rhetorically. Fair question, Jack thought. "It is because of *gravitational radiation*," continued the voice from the speaker. "Just as a whirling electric charge emits radio waves, so the theory of General Relativity predicts that a body with mass that is undergoing violent acceleration will emit gravitational waves. These waves carry away energy, and this energy is lost from the orbiting neutron star. Losing energy would normally tend to slow its motion, but because it is in an orbit the consequence of losing energy is that it moves into a closer orbit where it travels round more quickly and its orbital period will decrease. That is what we do see."

Yet another frame opened with a numerical display that gave the period of each successive revolution, and with the passage of time this became steadily less. A graph showed the way that the period changed with time.

"In this graph you see that the period of the orbit is steadily decreasing. The observed times are compared with the prediction given by General Relativity for the rate at which gravitational radiation will carry away energy, and you can see that the observation and prediction agree to within one percent, which is as accurately as we may measure the loss rate." Jack wasn't sure that he *could* so easily see that from the columns of figures, but he was ready to believe the Hubblegoblin. "This is an example of how careful measurement may be compared with prediction, and how accurate numbers make a convincing case, a lesson I learned from my cousin the Techno Troll," he finished.

"However, it is not only by the slow accumulation of accurate numbers that we may see striking effects. As I mentioned, we have reason to believe that giant black holes, such as might produce strong gravitational lenses, do exist in the centers of many galaxies, perhaps even of most. They might make their presence felt by the bending of light that passes them, as we saw earlier, and by their effect on the motions of stars close to them. It is seldom that the alignment is suitable to give an image focused at the Earth, and in general galaxies are so far away that it is not easy to observe how their stars move. There is another effect, though, that *is* dramatic enough to be seen clearly from a long way away. That is the appearance of an Active Galactic Nucleus: a *quasar*." The voice from the loudspeaker paused dramatically.

"The name 'quasar' is a shortened form of a *quasi stellar* object: an object in the centre of a galaxy that looks to be similar in size to a star but that may give enough light to outshine the billions of stars in the galaxy. Such objects are truly dramatic and would be uncomfortable neighbors if you were close to one. The ones that you may observe are all a long way away, however. They are very distant in space and that means they are very distant in time also. They are visions from the far past when the Universe, and galaxies were much younger than now. You can see a picture of one here."

The image in the view screen changed to show the curdled-milk brilliance of a distant galaxy. There were a few bright stars in the foreground, bright because they were so very much closer, but in the center of the galaxy shone a pinpoint of light, seeming as bright as the foreground stars despite its enormously greater distance from the observer. When this was pointed out to Jack, he asked how one could be sure that the object was in fact far away and not a star that just happened to be in front of the galaxy.

"I'm glad you asked that," broke in the Intellectual Giant. He did appear to be glad and probably was, since it gave him an occasion to get back into the discussion. "It is best seen through the Hubble expansion of the Universe," he began. "It has been observed that the galaxies in the Universe are all flying away from one another. This is not simply because the galaxies happen to be moving *through* space and getting ever further apart. No, it is because space itself is expanding. Now that gravity may be seen as a curvature in SpaceTime, it becomes clear that space and time are not an unchanging backdrop to the action of the Universe, but active participants in the drama. With space and time combined it becomes most reasonable that space may change with time; either it will expand or it will contract. An eternally constant and unchanging space is improbable, like a pin balanced on its point. It is almost certain that it will fall one way or another and it has chosen the option of expanding, of blowing up like a balloon. As space increases so the galaxies all become farther apart.

"The speed at which a galaxy recedes from us is approximately proportional to its distance, and you can find this from the *red shift* of the light from it.[12] Light from any material objects

. . . .

12   In *"Ali Gori and the Cave of Night"* we spoke of red shifts as a measure of velocity and of the expansion of the universe.

shows narrow lines in its spectrum in patterns that are unique to the chemical elements it contains. From distant objects the pattern of lines is unchanged, but they are all shifted toward the red end of the spectrum. This is a measure of the speed at which they rush away, and the greater the shift, the greater the distance. In these galaxies with active nuclei, the light from the stars and that from the quasar in the centre show the same red shifts and may be taken to be at the same distance. We may reasonably say that the quasar *is* in the galaxy.

"Further support for this comes from the long *jets* that quite frequently appear on either side of the active core." The speaker drew attention to two thread-like features on the viewscreen that started close to the bright center of the galaxy and protruded for some thousands of light years on either side before ending in diffuse blobs.

"From the Doppler shift[13] of the light from these jets you find that they are fast-moving gas. It is moving very fast, almost at the speed of light. Sometimes you see that a jet moving toward you appears to be *superluminal.* If you measure the distance that irregularities in such a jet have moved in a given time, you may find that the jet appears to move faster than the speed of light. It doesn't actually do this of course. Because the jet *is* moving very fast, it does go a long way in the time that you watch it. Its end point is a lot closer than it was when you began to time it, and so the light from it does not take so long to reach you as the light from the start. This means that the time you observe between start and finish appears too short and so it appears to move unreasonably fast.[14] You always have to remember that what you see is in the past to a greater or lesser degree. You never see the present." He paused and took a breath before returning to his main topic.

"It seems that these Active Galactic Nuclei are in fact black holes. That must strike you as strange, since a black hole's most obvious feature is that nothing can escape from it, neither light nor anything else. As seen from the outside, which is the only way you can see one, a black hole is an object of remarkable simplicity: mass, rotation, and possibly an electric charge. It has no other fea-

* * * * * * * *

13 This was mentioned in 'The Prince and p'. The light from something moving rapidly toward you is shifted toward the blue end of the spectrum, toward the red if it is moving away.

14 This is rather like the Blue Knight's lance in 'The Prince and p'.

tures. This bare simplicity is expressed by saying that 'a black hole has no hair'. Maybe it does not, but it may wear a wig. To help you imagine what I mean, you had better follow me."

The Intellectual Giant led Jack through a door that had not previously been visible in the wall of the turret room, and the pair found themselves again in the control room of the starship. "We didn't come through that door the last time," protested Jack. "No, it doesn't matter though. This is not really anywhere so it doesn't matter how we got here."

The ship had obviously traveled quite some distance since they were last aboard. "Clearly a different episode," Jack's guide muttered under his breath. As before the crew were all staring at the large forward view screen that showed a great swirl of gas, glowing with a blinding brightness. From the center of this emerged two opposing jets that extended clear out of the picture on either side.

"A large black hole exerts an enormous gravitational pull on its surroundings, and if there is matter nearby, it will be pulled into the hole. In general the matter will initially have had some sideways movement, so that it cannot fall directly into the hole. It has angular momentum and so goes into an orbit and adds to an *accretion disk* that swirls around at some distance from the event horizon, rather like a cosmic whirlpool. As the molecules of gas course around, they collide with one another and may lose energy and gradually spiral into the hole. What does this have to do with the brilliance of galactic nuclei, you might well ask?" Jack admitted to himself that he might well, but as the Intellectual Giant was obviously going to answer anyway, there did not seem much point in asking. Surprisingly he chose not to answer himself. "Let us stand here a moment and you will hear the Astrogator explain what you are seeing."

"What is that thing!" asked the Captain, in the voice of one who is decisive if rather poorly informed."

"It is the accretion disk around a giant black hole, Captain, said the Astrogator with the assurance of a digital library. "Gravity around the hole is so strong that anything falling toward it will accelerate to a speed very close to that of light. The impact of gas and other material moving at such a speed on the accretion disk heats the material in the disk up to a very high temperature, a temperature quite outside our experience, Captain." Interstellar gas from within the galaxy drifted down like a gentle fall of snowflakes, gradually accelerating toward whatever hell awaited it near the black hole. "This means the disk glows, and it is receiving so much energy, it glows more brightly than all the billion stars in this galaxy shining together."

"I cannot believe that," snorted another officer who was lying on a couch to one side.

"You should believe it, doctor," responded the Astrogator calmly. "The strength of gravity involved is enormous. When a rock falls to Earth, it may build up kinetic energy that is equivalent to only one billionth of its rest mass energy, and yet it appears that it was an asteroid strike that destroyed the dinosaurs and left darkness covering the Earth for generations from the dust flung into the air.

"When it reaches the event horizon of a large black hole, an infalling object may have a kinetic energy that is as great as its rest mass energy. When such an entity collides with dust and other grains in the tangled maelstrom of the accretion disk, it may emit around ten percent of its rest mass energy as radiation back to the outside world before it plunges to oblivion. Compare this with the light from a star. The nuclear processes within a star may release about half of one percent of the mass energy of some of the material in a star, and it does this typically over a period of billions of years. Material that falls into the accretion disk of a black hole emits much more of its energy and quite quickly. As you can see, a large hole can devour a whole star."

A verse came unbidden and somewhat modified into Jack's mind.

> The cataract of the cliff of heaven fell blinding off the brink,
> The dust of heaven came roaring down for this throat of hell to drink
> And even washed the stars away as suds go down the sink.
> The Astrogator cocked his eye "An accretion disk I think."

On the screen, which was now showing a closer view of the swirl of hot dust around the hole, the glowing disk of a star could be seen drifting toward it and then moving ever faster towards its fate. As it closed to its destination, the star's shape blurred and elongated, and long streamers of hot gas peeled off ahead of it.

"That is the tidal effect. It is tearing the star apart as it pulls it in. Tidal effects come about because the strength of gravity varies with distance from its source. On the Earth the gravitational pull of the Moon varies noticeably over the distance from one side of the Earth to the other, so that the water in the oceans closest to the Moon is pulled more strongly than that on the far side. This gives a slightly elongated egg shape to the surface of the ocean and so there are bulges in it on either side of the Earth that follow the Moon around and create the tides. It is the same here, save that gravity is so much stronger. The pull on the nearer side of the star is much greater than on the farther side and so gravity rips the star apart as it approaches.

"You are seeing a star, which might otherwise have shone for billions of years, being absorbed very quickly. The rate of energy emission is enormous, Captain. It is the most powerful energy source known in the Universe."

"I thought that the matter-antimatter annihilation in our engines was more effective," said the Captain.

"Well yes, Captain, so it is in theory..." "Ye cannae get the anti-matter," interrupted the Engineering Officer's voice from a speaker. "Ye just cannae get it. It would be great stuff if we could get it ye ken, but there is just nowhere we can get it. I've been meaning to tell ye that Captain," he added.

"Right," said the Captain rather dubiously. "What about those jets though? Surely they are not something that you would expect from a black hole. Aren't black holes without exception round and featureless and allow nothing to escape, certainly not a huge jet of high speed particles!"

"Well, yes and no," replied the Astrogator in inhumanly calm, measured tones. "The black hole contributes only an enormous

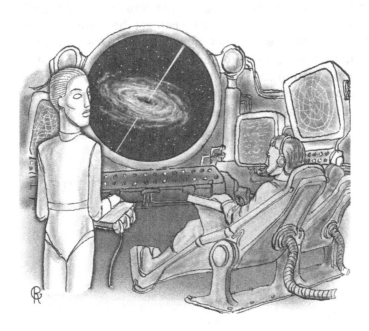

gravitational field–indeed, it *is* an enormous gravitational field. But that is all that is required from it. It is this that results in the orbiting high-energy dust and gas that make up the accretion disk, and it is the disk that emits the radiation and generates the jets."

There was a note in his voice that they had never heard before. On reflection they realized that it was *uncertainty*. He was in the uncomfortable position of expressing a hypothesis rather than established fact.

"With such high energies the gas is strongly ionized–the atoms have lost their electrons and so the material is electrically charged. A rapidly circulating electric charge is in essence an electromagnet, and the magnetic field acts to guide charged particles in two great jets, such as the ones we see just in front of us. It is rather similar to the way that the magnetic field of the Earth will guide charged particles down onto the north and south poles to create aurora lights." For someone who habitually travelled across the width of the Galaxy, his analogies were remarkably parochial. "In this case both the energies of the particles and the strength of the magnetic field are far stronger, so we get these enormous jets instead of a local aurora."

"Warning! Warning!" cried a mechanical voice, blaring through the control room. "Imminent collision with high energy particle beam!"

The Captain spoke swiftly into a microphone. "Emergency power to main engines, we must alter course immediately!"

"I canna do it captain. We no can get the antiparticles you see. We're going to dive straight into yon beam. We're doomed!"

In this dramatic moment Jack though he heard a slight whirring behind him and turned to see a camera aimed at the group on the bridge, while between him and its lens he saw letters scrolling upward. As he read them from behind, they said

Continued in next episode

INTRODUCTION TO

the fifth tale

"So now I have told you something about the Universe, about the stars and galaxies it contains and also how the fabric of space and time is distorted to provide the gravity that holds them together," remarked the Storyteller. "I have tried to give you some sort of picture, however inadequate, of what the Universe around us is like. The time has come to talk, not of how we see the Universe, or even of how it really is, but of how it came about. Whence came this Universe of stars and galaxies, of dust clouds and background radiation?

"I cannot really answer that question, but I can tell you something of the stages along the way, and some of them were very close to the beginning of the journey. Our Universe has had a long history, at least by any human standard, and at times a rather lively one. so now I shall at last tell you about the Big Bang.

"Sit quietly and I shall tell you the story of 'Waking Beauty.'"

"About time, too!" muttered Adam rebelliously.

*The Storyteller chose to ignore him and instead folded his
hands neatly in his lap and began his tale.*

*"There was in a remote kingdom a king and queen, who
were blessed with a baby daughter. To her christening they
invited all the nearby royalty and nobility. They also invited
the good fairies, who blessed their daughter's nativity with
the gifts of beauty, of charm and of intelligence. The parents
had neglected to invite the local wicked witch, but unfortunately she came anyway....*

the fifth tale:

WAKING
beauty

(the big bang and after)

The witch looked down upon the baby Aurora in her crib. "You are a lucky little girl," she crooned. "Everyone is giving you the most wonderful presents and you will grow up pretty and clever. Clearly, you must be destined for a wonderful future—or are you? Enjoy your life, for it shall indeed be blessed, at least for the first sixteen years." She paused in a sinister way. "When the clock strikes on your sixteenth birthday, it shall strike *you* and then you will find that you *have* no future. He, he, he. He, he, he, he, HEE!" Her voice rose higher and higher in an exultant cackle. Then she looked around and saw that everyone was looking at her with an expression of polite bewilderment. "Well, that's it, then!" she snapped, and vanished.

When he heard the witch's curse, the king immediately decreed that all clocks should be banished from the kingdom. Initially he set up a select group drawn from members of the City Watch, who would stand on high towers and at upper-story windows and shout out the time as loudly as they could. Unfortunately, as none of them had any clocks either, their calls got steadily out of synchronism with one another and members of the populace complained that the constant shouting was giving them a headache, so the scheme had to be stopped....

As a consequence, no one in the kingdom was ever really sure what time it was. The peasants just had to get up in the morning when it was light, stop working for meals when they were hungry, and go back to bed when it got dark. Mind you, this was what they had always done, as they had never had clocks anyway. The merchant class had to make more of a change in their habits. The first of them to wake in the morning sent a servant to run round the town waking all the other tradesmen, and the morning was then officially held to have begun. There was the advantage that no one was ever late for an appointment, since they could not make any. The main problems arose in the palace itself. The palace had been full of clocks and royal routine had run with literally clockwork precision. The royal entourage managed as best they could, but it was generally frustrating.

On the morning of Princess Aurora's sixteenth birthday, the royal family was sitting around the table having breakfast. At any rate, most of them were having breakfast. Cousin Beatrice was under the impression that it was a light lunch and Uncle Alphonse, who did not get out very much, thought he was having a very late supper. On the whole, though, they were agreed that it was breakfast time.

As her family discussed her forthcoming birthday celebrations, with a certain amount of bickering, Aurora felt an urge to slip away for a little while and wandered off around the castle. Eventually, she found herself climbing a high stair that she did not clearly remember having seen before. At the top there was a door that opened into a small attic room, and in the corner stood a tall rectangular case. It was a strange thing. At the top there was a sort of circular face with numbers around the edge and two pointers fanning out from the center of the circle. Lower down there was a glass-paneled door behind which a large weight on a rod swung ponderously to-and-fro.

Aurora was fascinated. She went up this intriguing object and opened the door to have a closer look. As she did, the pendulum, for such it was, paused in its measured motion. It swung abruptly out and struck the girl squarely in the chest so that, as predicted, Aurora found that she had no more future. That is not to say that she died, but only that for her the normal passage of time no longer held. As if in a dream she saw herself reversing away from the clock and walking backward out the door. She saw no more of the little room, for her perception fled out the

window and soared up into the sky, rising above the clouds and out into space. Time, or whatever now passed for it, had speeded up so that she saw the Earth turning on its axis below her and then she saw the Moon swinging in orbit around it. Both appeared to be moving in the wrong direction, but she didn't know that.

Her perception of her surroundings widened ever further. Sundry planets swam past her view and then disappeared in the vastness of space. As she soared farther, the whole system of the Sun's family shrank to a bright dot and other stars came into view, more and more and more of them. Soon, or whatever word is appropriate in the absence of properly functioning time, she saw the great disk of our galaxy and it, too, could be seen to turn. Other galaxies were visible, a great multitude of them. It became apparent that they were moving toward one another, each one crowding steadily in upon its neighbors.

Aurora was observing the galactic expansion in reverse. As the galaxies drew together she saw changes in them. They developed a greater preponderance of blue-white stars, blazing brightly, if briefly, in the young days of galactic life. Many of the galaxies showed intense cores that outshone the light from all the stars present. The Universe was a bright place with brilliant fires new-kindled in those times. As she observed this reverse history of time, the stellar fires blinked out as they went through the reverse birth pangs of condensing clumps of gas, observed now before their nuclear ignition. The Universe was totally dark to her sight.

There was a brief period of utter confusion. Dramatic events took place, but they were too swift and confusing to mean anything to her. Then she was–elsewhere.

"Right gentlemen!" said a rich and rather self-satisfied voice. "As the main representative of the Company, it is my pleasant duty to address all you prospective stockholders in this outstanding enterprise. This is an opportunity that does not come along every day, or even every aeon. You may all congratulate yourselves on getting in on the ground floor, so to speak, of this dramatic new venture, a project to be known as *The Universe.* You have been invited here, as the representatives of many of the most respected creation myths, because the board felt you would have a special interest in the project. If you would all sign your copies of the document of incorporation we may begin the process, which has been amusingly dubbed by our PR people as 'The Big Bang.'"

The speaker was somehow hidden from Aurora's sight, so she looked around at his audience instead. They were a large and rather motley collection. There was a preponderance of tall figures in shining robes, with and without wings. There was a compact group of heavily armed and bearded gods and ice giants, and also sundry huge birds and tortoises, and many interesting shamans and minor deities. One particularly hairy figure with a huge axe complained that it wasn't like this in the old days. When another

member of the group pointed out that, as time had not yet begun, there could hardly have been any days older than this, he sulked and would not say another word.[1]

"And so, ladies and gentlemen and sundry mythical creatures," the voice boomed out confidently, adding "and various body parts " as he noticed the presence of the Great Nose whose First Sneeze had sprayed the stars across the heavens. It was fortunately a *very* minor shareholder in the project, but its presence still deserved to be acknowledged. "And so," he continued, "as chairman of *Universal Universes Inc.* I give you–the Universe!"

In front of the assembled watchers appeared a small, intense dot. It would not be true to say that it shone brilliantly since light did not exist at this stage, but it was still in some way perceptibly intense. It was *The All.*

Hardly had Aurora been able to register its presence than the dot suddenly expanded. "Expand" is indeed far too weak a word to describe its behavior since it exploded from a tiny speck to a vast area that seemed to be as large as the Universe. Indeed it *was* the visible Universe at that time.[2]

"What has happened?" the princess wondered aloud.

"That was the inflationary phase of the Universe: a period of exponential expansion," a voice answered her. She looked around and saw, floating beside her, a young man in a neat suit whom she recognized as one of several "associates" who had been standing to one side at the meeting. He wore a lapel badge that read "I'm Paul, I try harder." He was carrying a clipboard on which there were several sheets of paper. She could see that the top one was titled "The four main problems of the Universe." She thought in passing that this did not seem all that very many.

"We have left the Planck epoch," stated her companion. "That was the initial stage where gravity was so intense that the very fabric of SpaceTime was governed by quantum physics. Time and space were not smooth and continuous there, but split into little lumps in a sort of SpaceTime foam, or so I am told.

• • • • • • • • •

1 LEGAL NOTICE: The management wishes to point out that time is solely a property of the Universe™. Any words used that might suggest a chronological or temporal relationship are not to be construed as any form of warranty as to the existence or effective presence of time at this time. (Sorry, the word "time" just slipped out there!)

2 It's OK. I can say "time" now as we are in the developing Universe.

Gravity was so strong that time was bent to become space-like. The time and space axes of SpaceTime were no longer distinguishable and had flowed into one another, so that there was no clear beginning of time, let alone a past.[3] Time is embedded in the Universe and so had its beginning as the Universe began. Fortunately, this stage did not last very long, probably less than a million billion, billion, billion, billionth of a second."

"It sounds totally confusing," said the girl.

"Oh it is. I cannot make head nor tail of it myself. We leave that sort of thing to our backroom experts. I have enough difficulty trying to sort out these problems." He waved his clipboard at her. "It is all right for the management. They make the broad executive decision and then leave it to their staff to make it work out. I and the likes of me have to fill in the details. I am not entirely sure about some of the early bits, but it seems to get there in the end. Look at this for example." He held the board out to her and indicated the top item. It read "The Horizon Problem."

"That arises because the Universe, when it is observed later on, is seen to be very uniform on the large scale. If you look at the visible Universe in your time, you find that, apart from little local irregularities like stars and planets and galaxies, on the really large scale one bit is very much like the rest. Now if something is much the same everywhere then it is usually a fair guess that it has been mixed together at some time. If you have a boiling drink and a handful of ice cubes and you put them together, then if you wait for a while, they will get mixed up to give something that is neither boiling nor freezing but much the same everywhere. In such mixing all the big variations will have smoothed out, but it takes time. The problem," he went on, emphatically shaking his clipboard, "is that there seems to not have been enough time for it to happen.

"By tracing back the observed Hubble expansion until you see everything come together at one place, you can arrive at a rough age for the Universe. You find there has not been enough time for a light signal to travel across from one extreme of your visible Universe to the opposite side, let alone for the many exchanges you would need for any sort of smoothing process. The Universe has been expanding continuously of course, so in the past things were

· · · · · · · ·

3 Such a condition is inherently confusing. The absence of a past is often compared to the fact that you cannot go farther north from the North Pole. This may not be a very helpful analogy, but it does make one think.

closer together. However, the Universe had not existed for so long and so the *visible* Universe was smaller because light moves at a limited speed. As a general rile, the more time passes, the more of the distant Universe comes into sight, as there has been more time for the light from it to travel to you. Much of what is now your visible Universe would have been unable to exchange light signals in the past."

"Are light signals the only option, though?" asked Aurora.

"Perhaps not, but nothing can travel through space faster than light.[4] We have a way round this problem though. We do it with inflation."

"What is this inflation and how does it help?" Aurora had been brought up to speak carefully and organize her thoughts.

"Well," began her companion, "we start with the false vacuum." Aurora wisely did not say anything as she suspected an explanation of sorts was to follow. "The false vacuum was a condition where all was simple and symmetrical. There were no different forces in that they are all aspects of the same basic force and at that time they were indistinguishable. All was subsumed into the one. It was a time before particles, and the gravity of 'empty' space was all that existed.

"In the false vacuum there was a gravitational force of sorts, and it was in effect repulsive on the SpaceTime frame of the vacuum itself. There were no particles at this stage, and in the absence of an attractive gravity between particles that might balance this repulsion, space expanded. Particles that feel repulsion undergo a finite acceleration, that is limited by their mass. As there was nothing equivalent to the mass of particles, space expanded rather rapidly–very rapidly in fact. The overall result was that, in less than a billionth of a second, the space had doubled in size some ten billion times. That is enough to expand a hydrogen atom to a size vastly greater than the extent of the Universe visible in your time. This means that before inflation happened, the region that is now your visible Universe was quite tiny and could mix. I thought of that,"[5] he added proudly.

"It seems rather extreme," said the princess judiciously.

"Well, awkward situations call for extreme measures," answered the young man smugly.

• • • • • • • •

4 This is required by Einstein's relativity theory. OK, it was a *theory* for Einstein, but it has been very well proven since he proposed it.

5 No he didn't. It was in fact Alan Guth who first suggested the notion of cosmic inflation.

"But doesn't all this expansion mean that things are moving faster than the speed of light?" said Aurora.

"Oh no, not as such. Nothing material can move through space faster than light and no information may be transferred any faster, but we are not talking about anything moving *through* space. This is space itself that is expanding, and that is quite different.

"Inflation helps with the next problem also." He turned over another page on his clipboard. The next heading was "The Flatness Problem."

"This is also sometimes called 'The Fine Tuning problem.' It relates to the amount of matter in the Universe and whether its gravitational attraction will be able to overcome the Hubble expansion. If gravity wins, then the expansion would slow to a halt and the Universe collapse again toward a final *Big Crunch*. It is what you might call an alpha and omega situation. As we are close to the very birth of the Universe this is certainly an alpha moment. The final condition of the Universe depends on omega, Ω–the ratio of the amount of matter that is actually present to the amount that would be needed to make the expansion of Universe slow to a halt. If Ω is greater than one then the Universe will eventually collapse on itself, and in this case SpaceTime is positively curved. Space and time would close in on themselves, a bit like the surface of a sphere. If Ω is less than one, then the Universe will go on expanding forever. If Ω were exactly equal to one, then the expansion would slow toward a halt but would take forever to do it. The expansion speed would 'asymptotically approach zero' as our backroom boys say. In this case SpaceTime would be flat and not curved overall. Hence the name "Flatness Problem."

"Wise men and scanners of the skies will find, in years that follow soon after your own time, that the total mass of all the visible stars falls well below what would be needed to halt the expansion of the Universe. But there are signs of other mass that gives no light and cannot be seen in their telescopes. This 'dark matter' is known from the visible effects that its gravity has upon the stars orbiting within the galaxies. These orbits would require more gravity in the galaxy than would be produced by the mass of the visible stars.

"The total matter present is found to be at least a tenth of that needed to make Ω equal to one–possibly even being the full amount required. It is rather difficult to tell when you cannot see most of it. Such close equivalence would require that the Universe begin its long history with expansion and matter remarkably balanced."

"One thing ten times another does not sound like a fine balance," objected Aurora, who had been brought up to be clear about such things.

"Ah, but that factor of ten is after the passage of a great deal of time. All of the galaxies are rushing away from one another's gravitational pull. It might help to see the effect of this if we thought instead of something leaving the gravitational pull of the Earth.

"Imagine, if you can, that you lived in such a time that there were craft that could rise from the Earth and sail into the heavens, actual ships of space. In order to escape from the Earth they must be projected from the Earth's surface with at least the *escape velocity*. This is the velocity that would give a kinetic energy as great as the gravitational binding energy with which the ship was tethered by the Earth's gravity. Less energy than this and the ship will falter in its rise and fall back to Earth. More energy, and it will leave with a surplus of kinetic energy that it will carry with it far beyond your sphere. If the initial speed of the ship was exactly equal to the escape velocity then the ship *would* leave, but its speed would fall lower and lower. It would have no surplus of kinetic energy, so that at a very great distance it would but barely be moving away from the Earth.

"Now you may clearly see," he continued hopefully, " that should there be *any* significant residual speed, then as the speed *needed* to escape any further falls toward zero the ratio of the finite actual speed to this steadily decreasing required value will become greater and greater. Anything, however small, is greater than zero by a large factor. As the ship travels ever farther from the Earth, so the Earth would need a larger and larger mass to pull it back again. The mass of the Earth would be too small to hold the ship, and that by an ever-increasing factor.

"Much the same considerations hold for the expansion of the galaxies. Let us assume that the initial velocity of cosmic expansion was almost, but not quite, balanced by gravity. After much time had passed you would still expect to find that the Universe contained far too little mass to halt the expansion. The observed agreement to better than a factor of ten is quite remarkable.

"Cosmic inflation would predict a value of Ω very close to one, with SpaceTime flat overall. Inflation tends to smooth out any curvature of SpaceTime, in the same way that a huge balloon has a less curved surface than a smaller one. A value of Ω equal to one is equivalent to a space ship with precisely the escape velocity from Earth, where kinetic energy and gravitational binding energy are exactly equal and opposite, and the total energy is zero."

Paul turned toward Aurora and held up one finger to emphasize his point. "If the Universe started from something very minute with no overall energy, from some infinitesimal seed that inflated into the present Universe, then you might expect the total energy to be zero. The positive energy contained in the motion of matter within the Universe would be exactly balanced by the negative energy of gravitational binding. This would give a Universe with a minimal initial capital outlay of energy–a very attractive feature for our management. This corresponds  to a value of Ω equal to one. In this favored scenario, cosmic expansion will probably continue forever, though slowing as it coasts toward a halt that will never be achieved."

"You must bear in mind that we require a Universe of Life," cut in a new voice authoritatively. Aurora turned to see a figure in a long flowing robe and almost as long a flowing beard. He was floating nearby in the false vacuum of the inflated Universe and holding in front of him, not a tablet of stone, but a copy of the contract they had all signed.

"I have been chosen as representative of the stockholders and this document explicitly states that we require a Universe that contains life. If there were too much matter then the Universe would collapse again before there was time for stars to ignite and illuminate planets over the long periods required for life to develop. Contrariwise, if there were too little matter then the expansion would be so fast that galaxies and stars could not form. It is all in the contract."[6]

• • • • • • • •

6 I do not seriously suggest that the Universe must contain life because this has been stipulated in some initial contract. The *Cosmological Anthropic Principle,* should you believe in it, does however state quite seriously that the Universe we see must have been carefully adjusted so as to create life, after all we *are* here to see it. Either it was somehow designed that way, or out of many possible universes we are seeing the one that happens to be capable of producing life.

. . . .

The Cosmological Anthropic Principle

This is a notion that seeks to explain the fact that the Universe seems to be amazingly well suited to us. If the values of some of the many very basic quantities had been slightly different, then the Universe would not have been able to develop our sort of life. If the weak interaction had been just a little stronger, the stars would burn out too quickly for life to develop. If gravity were weaker galaxies and stars would not form at all.

The *Strong Anthropic Principle* says that the very existence of a Universe *needs* minds to comprehend it (us) and so must be able to support life.

There is a weaker version that says we *are* here so obviously it must be such that we can live in it. You cannot really argue with that.

. . . .

"Have no fear, sir," responded Paul soothingly. "You may rest assured that we shall honor our contract in every detail. There shall be life. There shall be life abundant. You will just have to wait for a little time.

"But enough of this chatting!" he said turning to Aurora. "We have been talking long enough and time will not wait for us.[7] As the inflationary period only lasts for a billionth of a second or so, we had better get.on to the next stage. There is a change of phase as the false vacuum changes to the normal vacuum of space. Normal space condenses out in bubbles, much as freezing water will change to ice in little specks and crystals rather than freezing all of a piece at every point. As the false vacuum changes, it releases its enormous energy to create particles, specifically photons. 'Let there be light!' you might cry, but it is light of an energy so searing as to be inconceivable to us." As he spoke photons were produced all around them in a cosmic fireball as the nature of the vacuum changed to its more familiar form. The false vacuum released its huge energy in a torrent of photons, photons of such energy that they in turn created particle-antiparticle pairs in great profusion. It was a period of democratic confusion in which all possible particles made an appearance, in each case together with their shadow companions–their antiparticles. In essence the world became a seething cauldron of quark soup, with electrons and other leptons

. . . .

7 It never does.

as spice to the mix. There was no problem with the energy needed to create them, since energy was available in abundance.

"One of my problems arises at this stage," continued Paul. "A photon will always create a quark *and an antiquark*. The case is similar with the electrons that will eventually go into making atoms. Each of these is always created together with its antiparticle, the positron. There should be as much *antimatter* around in the later Universe as there is matter and clearly there isn't."

"Isn't there? How can you be so sure?"

"Well, if there were antimatter around and it met up with normal matter then they would *annihilate* each other–the particle-antiparticle pairs would convert back to photons. These photons would carry off the rest mass energy of the particles, an energy that is both large and specific. This energy of *annihilation radiation* is clearly recognizable and you just do not see it. There is enough material in interstellar space to give detectable radiation by interacting with any of the stars in our galaxy that might have been formed from antimatter. If there is any antimatter in the Universe, it must be a long way away from the matter that is near us, and it would be difficult to see how matter and antimatter had managed to become so well separated. If quarks and their antiquarks had been present in exactly equal numbers, then you may predict that, as the Universe continued to expand and cool, they should mutually annihilate and convert to photons. They should almost all do this, with those few that manage to escape confrontation with their doppelgangers leaving a far smaller residue of matter, any kind of matter, than you see around you in your world.

"That is not acceptable," cut in a voice. It was the impressive figure of the stockholders representative again, brandishing his copy of the contract. "There must be matter," he thundered, while flashes of lightning played around his feet to add to the effect. "We must have matter. Matter matters. You cannot say that no matter is no matter since the very existence of matter is a material clause in the contract!" He shook the document again while his voice and indeed the very heavens trembled.

"I shall soon get tired of this chap," Paul muttered to Aurora. He then announced loudly, "Have no fear Sir. Matter will be provided. The next of my four problems deals with this." He held up his clipboard again and pointed to the heading: "Baryon Asymmetry Problem."

"Yes," said Aurora with a hint of royal impatience, "and what exactly is a baryon then?"

"Oh, baryons form the nuclei of atoms, and atoms of course make up the matter in the Universe. They are combinations of quarks and there are plenty of quarks around, created by high-energy photons, as we have seen. We have noted, though, that each quark is balanced by the existence of its nemesis, of its antiparticle. If we are to have a Universe of matter, then we require an asymmetry. There must be more of one type than of its opposite so that when they destroy one another, there is something left. This asymmetry does develop and if you attend patiently I shall explain how." This last remark was directed at the shareholder, who did not appear to be a patient type.

• •

Seven ages of the Universe

1 *The Planck Time.* The physics involved was strange, unfamiliar, and debatable. Energy levels were high and the period led to Inflation, when the Universe expanded in size beyond our conception.

2 *Particle creation.* The enormous energy in the false vacuum went into the creation of particle-antiparticle pairs. They are in a form of thermodynamic equilibrium and freely change between types. We deal now with familiar processes.

3 *The expansion of space increases wavelengths and lowers energy.* The particles decay to the particles of lowest mass: neutrons and protons. There is some tiny loss of antiparticles relative to particles.

4 *Nucleons combine to form nuclei.* Electrons are still free and scatter light, thus producing extensive thermodynamic mixing and smoothing out any irregularities in the distribution of matter and hence of the future Universe.

5 *Neutral atoms form.* There is now little photon interaction. Most photons remain as the cosmic background radiation, now ignoring the rest of the Universe as their energy decreases steadily.

6 *Galaxies form.* Clumps of dark matter hold the material together against the expansion of space. Galactic gas curdles into stars.

7 *Stars form.* Newer ones include heavier nuclei created in now dead Supergiant stars. Planets form around them with a sufficient fraction of heavy elements, and from these eventually build life—US!

• •

"Up to this stage the weak, the strong, and the electromagnetic interactions between the particles were still unified—they were not clearly different from one another. The Universe expanded at the beginning and has continued to expand throughout all of your past. As space expands the wavelengths of photons and other particles that inhabit this space are constantly stretching with it, and when their wavelengths increase, so the photon energies decrease, since a photon of longer wavelength is a lower-energy photon. The temperature of the Universe dropped. Temperature is just the energy available to each particle, and as this fell, the situation changed. The simple symmetry of the interactions that control the behavior of particles was broken, and there emerged the complex behavior that we are accustomed to seeing in particle physics. Particles gained mass, and in their masses they differed greatly. Heavier particles could decay to lighter ones and some of these decays were through the newly distinct weak interaction.

"Now I have to tell you that the weak interaction is a weird one. Though it is weak, it can do what other interactions may not: it can and does convert one type of quark to another. The weak interaction can do more than this. It can on occasion convert between particles and antiparticles. It does not do this easily or frequently; indeed it can hardly do it at all. Nonetheless, the time comes when there is a difference of about one part in a billion between the number of quarks and of their antiquarks. This may seem insignificant," he said equally to the two members of his audience, each of whom was showing some sign of impatience in his or her own way. "It is, however, vital to the appearance of a Universe composed only of one type of matter.

"Throughout all this activity the Universe continued to expand, as it did at the very beginning. Particle momenta may decrease with the expansion, but the *rest mass* energy of particles is unaffected. This latter is an intrinsic property of the particle and does not alter when space expands around it. Consequently, as the photon energies fall due to the expansion of space, a time comes when the photons no longer have enough energy to create the rest mass for new particle-antiparticle pairs. When protons or neutrons annihilate on meeting their antiparticles they create photons in pairs in order to conserve momentum, and now the photons so created do not individually have enough energy to create further-particle antiparticle pairs. The annihilation of the particles becomes a one-way process and soon the complementary pairs have almost entirely eliminated one another, and the Universe is a sea of photons with a few residual particles of matter, like needles in a haystack.[8] The antiparticles had been eradicated and only that *one in a billion* population of unmatched particles remained."

The princess watched with interest as the antiparticles were effectively eliminated from the Universe, taking with them virtually all the particles that had previously been their counterparts. As the last antiparticle annihilated, she noticed nearby the abrupt appearance of a cylinder of sparkling light that rapidly cleared to reveal a craggy-faced man in a space suit. He looked around him with a certain air of satisfaction.

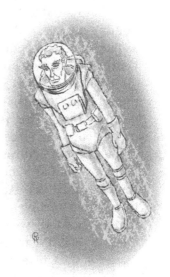

"Aye, 'tis just as I was telling the Captain. Ye cannae get the antimatter. Sure and it would be gae useful stuff if you could get it, but ye cannae. It is just not there to be had." He wore what could only be described as an expression of dour triumph as the sparkling cylinder formed around him again, and when it cleared he was gone.

"Now," went on Paul, totally ignoring this fictional interloper, "you see that you have here a photon-dominated Universe with but a sprinkling of matter in the form of nucleons that have no balancing antiparticles. That sprinkling is, however, sufficient to provide all the matter in your Universe. So, you see, the matter that you demand for the Universe has been provided[9]," he added, addressing the looming figure of the mythical stockholder.

Aurora abruptly became aware of two opposing groups that faced one another contentiously. On one side stood array upon array of tall shining figures, in inconceivable numbers stretching apparently to infinity. Opposing them were a few squat individuals, clutching to themselves a collection of rocks, trees, and consumer electronics, all manner of *things.*

"Behold the triumph of the Legion of Light!" chorused the shining shapes. "We are the representatives of photons everywhere, a Universe of light! See how we triumph over this sad rem-

---

8 Note that as this was the first time in the Universe that this expression had ever been used, it was, in fact, a vibrant and evocative new simile, spoiled only by the fact that neither needles nor haystacks existed at that time.

9 I do not really wish to suggest that these various requirements are all laid down in some contract drawn up at the dawn of time. However they all relate to what we actually see now, so they truly are requirements of any theory.

nant of miserable Materialists." The crouching Materialists said nothing but grasped their comforting objects all the more closely as they waited hopefully for such things to become a reality in the Universe around them.

"Matter is now present in the Universe," continued Paul, unperturbed by this display, "and it begins to move toward a more familiar form. This is the epoch of the strong interaction, and quarks combine three at a time to form the more familiar protons and neutrons."[10] Aurora muttered under her breath that they were no more familiar to her than were the quarks, but the lecture continued regardless. "At this time appear the whole zoo of strongly interacting particles, and perhaps a selection of heavy particles that interact only weakly. These Weakly Interacting Massive Particles, or WIMPs, may be an important component in the dark matter of the Universe.

"As the general level of energy available continues to fall, we are left with only the lightest of the baryons, the protons and neutrons. They no longer have sufficient energy when they collide to allow them to create the rest masses of the heavier particles. They may still convert to each other fairly readily through the 'weak' interaction. Protons may become neutrons, or neutrons protons, with the emission of a positron or electron as necessary to balance the electric charges. The general energy levels of the particles and photons that are present are still high enough that the energy needed to create the relatively tiny rest mass of an electron is not yet significant. An electron, or for that matter a positron, is only two-thousandth of the mass of a proton.

"However, the Universe continues to expand as always and the energy of photons and particles continues to fall. Soon there is not even the energy to create the mass of a positron, which must be produced when a proton changes to a neutron. The reverse process, however, may still happen because the rest mass of neutrons is just sufficiently greater than that of protons that there is enough energy to create the mass of an electron, with a little to spare. Neutrons may decay to protons in free space, but there is only barely enough energy available and the decay is a slow process. The neutron has a typical decay time of around fifteen

· · · · · · · · ·

10 See the tale "Snow White and the Particularly Little People" and my earlier book *The Wizard of Quarks.*

minutes. This may not seem slow compared with the vast age of the Universe, but these things are relative. It is very long compared to the decay time of most particles. More to the point, it is long compared to the time scale of events just before the formation of nuclei.

"As the Universe continues to cool, neutrons can and do decay to protons, but this does not go on indefinitely. Eventually the energies of the protons and neutrons become so low that they become bound to one another when they collide, rather than simply bouncing off. Nuclei begin to form. At first deuterium is formed, this being one proton and one neutron combined together. The deuterium nuclei soon combine in their turn to form the much more stable nucleus of helium. This is pretty much as far as it goes. The helium nuclei cannot combine in their turn because it so happens that there is no sufficiently stable nucleus containing eight nucleons that they might produce. Any product would be so unstable that it would fall apart before it could be built up any further and the process of nucleus formation effectively stops. It is just barely possible to produce minute amounts of lithium, which contains three protons and four neutrons. The formation of nuclei comes about one hundred seconds after the beginning of time and at its end we have a Universe that contains protons and helium nuclei in a ratio of seven to one, together with tiny amounts of deuterium and lithium. This ratio comes from a detailed calculation and agrees with what is observed in practice.

"The decay of the neutrons to protons stops now that the neutrons are bound into nuclei. Within the nuclei, as in any bound system, there are different states with differing energy levels for the constituent particles. As with the electrons in atomic states, the Pauli Exclusion Principle restricts the number of identical particles, such as protons and neutrons, that may go into each state in the nucleus. This operates separately for protons and for neutrons. Protons are identical to other protons, and neutrons to other neutrons, but a proton is different from a neutron and so the two types of particle independently fill the available energy levels. A neutron within a nucleus may be in a state that is lower in energy than the lowest state still available for a proton, and the difference in energy of the states may be much greater than the energy that is equivalent to the difference in neutron and proton masses. Neutron decay would not release energy and so it does not happen. The neutrons within the nucleus are now completely stable.

"The selection of atomic nuclei available in the early Universe is now established. The nuclei remain in their fixed pro-

portions while the Universe still inexorably expands and energies decrease. When the energy levels of photons and other particles became too low to create the masses of electrons and
positrons, the last continuing particle production process must
cease. Photons no longer have the energy to create electron
positron pairs. The existing electrons and positrons annihilate
and the photons produced remain photons. The electrons and the
positrons are gone from the Universe apart from a relatively tiny
few electrons that could find no anti-partners, as had earlier
been the case with the nucleons. Now the Universe consists dominantly of photons–of light."

"Rejoice!" abruptly chorused the Legions of Light. Paul gave
them an irritated glance but otherwise carried on as if they had not
interrupted him.

"Though the energy of light may still dominate the mass of the
Universe, the energy of individual photons is now small. They
dominate overall simply because there are so very many of them.
There are a good billion times as many photons as there are massive particles. Light and matter are still closely mixed, as they have
been since both appeared upon our cosmic stage. Photon production may have ceased and those that are present have energy that
is ever-shrinking, but the photons can still scatter readily by striking free electrons, and to a lesser degree from the positively
charged nuclei.

"In this scattering, light and matter exchange momentum and
energy to reach an equilibrium in which they are uniformly
mixed. Soon this scattering, too, will end. The photon energies fall
so far that light can no longer ionize and dissociate atoms. The
electrons bind themselves to the existing nuclei to form electrically
neutral atoms and there are no more free electrons. Because light
interacts with electric charge and we now have neutral atoms,
there is very little interaction.

This formation of neutral atoms does not all happen immediately, you understand. It takes some half million years, but it ends
with the background light no longer scattering freely from matter.
The light at this stage is like that emitted from a surface at a temperature of more than 3,000 degrees, like the visible surface of a
star. This is the time of Last Scattering. Hereafter the light scarcely
interacts again in the lifetime of the Universe. Light decouples
from matter and follows its own fading and independent path
thereafter. In later times the light could be captured by a star or
planet, but space is so large that most photons never meet one and

proceed undisturbed. The Universe has become transparent to most light, and on the whole matter no longer sees it. The Dark Ages have come for a time."

"Rejoice!" the Legion again sang in compound chorus. "We triumph over the drab followers of matter. So it is and so it shall be!"

"Perhaps," responded Paul, this time deigning to notice the interruption. "Perhaps it is so at this stage, but it is not so forever. The ascendancy of photons does end. Photon energy may still dominate the mass of the Universe at this stage, but photon energies are continuing to fall because the expansion of the Universe never stops. The rest mass energies of particles are, however, unchanged. The photon has no rest mass and so eventually the photon energies become so low that even the vast number of the photons does not allow them to dominate the mass energy of the Universe. It took some time for the photons to lose out, I agree—about ten thousand years—but at the end of this time the Universe was matter dominated. The gravity produced by matter now had the more significant effect on the behavior of the Universe than that of the pervading mist of photons that persists between the galaxies."

"Oh Woe, Woe," chorused the massed ranks of the legion of light. It could be seen already that they had so far shrunk as no longer to merit initial capital letters. They were steadily shriveling and contracting so as to be ever less noticeable. With a last despairing chorus they flung themselves upon the triumphant Materialists.

• •

Cosmic Background Radiation

The existence of the Cosmic Background Radiation is one of the great cosmological discoveries of the last century. Many people had noticed that there was an annoying background buzz that interfered slightly with high frequency radio communication. It didn't seem to come from anywhere in particular, and eventually two physicists at Bell Laboratories looked into it.

They found it was coming from outer space and uniformly from all directions, not just from our galaxy. It had a thermal frequency spectrum at just the right temperature to fit with the estimated age of the Universe and the temperature expected for photons whose wavelength had been stretching ever since they decoupled from matter.

• •

Around Aurora and her guide the light darkened to red and faded into invisibility. Paul flicked on a tiny penlight that he had extracted from a pocket and pointed at his clipboard. "Now we come to the last of my four problems: 'The Galaxy Seeding Problem'," he remarked. "The Universe now contains matter, mostly in the form of hydrogen gas, and there will be fluctuations in its density. As the gas is quite hot, all its atoms will be rushing about and there will be more of them in some regions than in others, simply by chance. There is a gravitational attraction between a gas atom and every other gas atom, and so atoms will be attracted to those regions where there is more than the average number. If the gas were static, then there would be a steadily increasing drift to the regions where there was already a higher density of gas, and stars and galaxies could form. Unfortunately for the formation of galaxies and ultimately stars and planets, the gas is continuously expanding, or at least being spread apart by the expanding space in which it is imbedded. This spreading inhibits any tentative condensation that might come from the first slight variations in density."

Aurora felt she could see around her the spreading cosmic gas, like a fog being dispersed ever wider by a gentle breeze. Here and there were denser wisps but these soon blew apart in the prevailing draft. How could anything develop?

"What we see in this case," remarked Paul "is the Power of the Dark Side."

"By that I mean that we are probably observing the effect of dark matter," continued Paul. "Clumps of matter of some type that has little or no interaction with photons or the other material in the Universe. It can give a large enough gravitational core to hold gas against the background expansion of space. We may not be too sure just what form dark matter takes, but we know it is there. The gravity it produces has clearly visible effects on the matter we *can* see, and there is a great deal of dark matter, at least ten times the matter that is visible to us."

"After its period of intense interaction with photons, the gas in the Universe would have been thoroughly mixed and would not have sufficient concentration anywhere to provide seeds from which galaxies may condense. When photons were scattering freely from the negatively charged electrons and the positively charged nuclei, there was a steady exchange of momentum between photons and particles in a process of thorough mixing that rather effectively smoothed out any clumps or irregularities left over from early times. The remarkable sameness that thereafter characterized the big picture throughout the Universe may still be seen in the remnant of light that fills space. This is amazingly uniform, no matter in which direction you look.

"Things may be different for dark matter. Any material that does not interact at all with the huge flux of light, or indeed all that much with normal matter, could keep any initial variations that there might have been between its density in one place and in another. Such dark matter would not be smoothed and mixed from frequent battering by photons since, being dark, it does not interact with the light."

• •

Dark Matter

There is evidence for dark matter in the Universe, matter that is not contained in the stars that we can see. The size of stars of different magnitudes is fairly well known, and you may estimate the mass of a galaxy. It would not provide sufficient gravity to keep the stars circling in their observed orbits within that galaxy.

If there is truly dark matter, matter that never does interact with light, then it would not have been uniformly mixed in the period of intense photon-particle interactions before neutral atoms formed, and might well be sufficiently lumpy to provide centers around which galaxies might condense.

• •

Aurora had a vision of nuclei and electrons being continuously shaken by their interactions with a vast sea of photons. Shaken, stirred and mixed into a smooth featureless distribution with any original irregularities in their distribution smoothed out by the severe agitation from the enveloping light. She was aware of dark background shapes lurking within this all pervasive activity, dark matter that was completely indifferent to the pounding sea of light because, being truly dark, it had no interaction with the photons. Unlike the electrons and nuclei, shaken and stirred to uniformity, this dark matter could retain whatever clustering it had from its first formation.

"Even though the dark matter may not have been affected by the photon scattering," continued Paul in his usual earnest fashion, " it can affect the more familiar forms of matter through the gravity it generates. Whatever the dark matter may be, it will have energy and mass and so will produce gravity. If it is clumped at some position, it can attract gas and so provide the nucleus for a galaxy. When enough gas has been attracted, the young galaxy will pull in yet more gas by its own gravity and the process is self-sustaining. The galaxies will grow until they have depleted the gas in their vicinity. In the process the density of the gas in the galaxy will increase and eventually it is sufficient to support the birth of stars. The time of starlight begins."

Aurora was aware that the gas filling the Universe was not everywhere expanding smoothly with the universal stretching of space. Here and there it snagged in the spider's webs of gravity that spread out from the scattered clumps of dark matter. The gas condensed in wide clouds. Clouds that kept their size and even condensed slightly, edging ever further apart as space continued to expand. Gravity was exerting its grip upon the new galaxies at their birth. The gas contracted toward the form of a sphere and, had gravity had its way unhindered, the cloud would collapse to an ever-smaller sphere.

However, the insistent pull of gravity was in competition with the need to maintain whatever angular momentum had been present in the original gas cloud. The gas had had some turbulent motion and this had resulted in a little rotation. Angular momentum depends not only on the mass and velocity of the particles but also on their distance out from whatever is the center of rotation. Although the velocities might have been small in the original extended cloud of gas, the distances were huge and so as the gas collapsed the relevant distances became smaller, and consequently the speed of the rotating gas had to increase. There was competition between the angular momentum, which wanted everything to be spread out, and gravity, which wanted to make everything small and compact.

After a lot of jostling and shouldering aside of gas molecules, a compromise was reached. Most of the gas continued to collapse smoothly to form a compact sphere, but a small amount of mass picked up a large velocity and swung around the center in a wide radius. It formed a disk that surrounded the more compact center and the galaxy took on the form of a bulging central core surrounded by a rotating disk.

Within the galactic cloud there was, as always, some turbulence in the gas. As the gas eddied to-and-fro some portions became briefly of higher density, some of lower. Or rather, the enhancements would have been brief were it not for the grasping

hand of gravity. Wherever the density was for the moment higher, then that region had greater mass and it pulled other gas toward it. Gravity pulled in yet more gas from the surrounding regions and the density increased. The galactic cloud curdled into myriad blobs of density greater than the average, and they grew ever more dense at the expense of their immediate environment. A billion stars took the first steps toward their fiery birth.

Each concentrated globule of gas drew in upon itself ever more rapidly under the contracting effect of gravity, and as each atom fell toward the center it released some gravitational energy. The gas heated up and in due course began to glow. The globule had become a protostar.

"Let there be light, as the saying goes," boomed a voice cheerfully from nearby. The princess turned to see a group of vast figures clustering around one of the gently glowing protostars. Many were clad in furs or shining robes and most in flowing beards. They bore a variety of battle-axes, flaming swords, and such. There were a few exceptions. There was a figure with an exotic headdress and a remarkably large number of arms. There was an remote figure in a linen skirt with the head of a hawk (in profile). There was a buxom lady in a toga-like garment with an elaborate hairdo whose aloof aspect was not in the least diminished by the fact she had an owl on one shoulder. The group was quaffing what appeared to be hot sweet tea from chipped enameled mugs.

"Who are you?" queried the princess in some surprise.

"Why, we are the early shift of demigods who have the task of overseeing this stage of the development of the cosmos," answered the one who had first spoken. "We are on the owl shift, as you might call it, begging your pardon Ma'am," he added with a nod of his head toward the matronly figure with the owl. "We have to be here to check how things are going. It's deity work, but someone has to do it."

"Oh," said Aurora, intrigued. "What do you actually do?"

"Oh, we mostly supervise. We do not actually have to do a great deal because things develop pretty smoothly on the whole. In fact," he added with a burst of candor, "we do not have to do *anything*, but don't tell anyone I said that."

"Why are you here, then?"

"Ah, that is because our local pantheons are very strongly unionized." He looked slightly embarrassed, as far as Aurora could tell from outside that expanse of beard. "If you like, I could point out some of the stages of star development," he added quickly, with the helpfulness of someone who has become totally bored by sitting doing nothing for millennia.

"This is a fairly typical star for this early period. As you see, it is contracting toward a denser center under the effect of gravity and in the process compressing the plentiful cloud of gas in the immediate neighborhood. When the gas falls into the compact center, it loses gravitational energy, d'you see, and this energy heats up the gas. You can see it glowing ever brighter and this build-up can go on as long as necessary," he added helpfully.

The princess felt that she was missing something here. "Necessary for what?" she asked.

"Necessary for hydrogen fusion, my dear," responded a figure in flowing robes, rather condescendingly she thought, as he took up the tale. He reminded her a little of the stockholders' representative whom she had met earlier. "The Universe is virtually all hydrogen and helium at the moment, and there is not too much we can do with that. We need to have heavier elements. We need stuff like carbon and oxygen and nitrogen: the sort of elements from which you yourself are built. Stars can make them for us," he added, "once the fusion reactions begin."

Aurora refrained from saying anything at this point. She was pretty sure she would soon be told all she wanted to know about fusion reactions, possibly even a bit more.

"In fusion reactions smaller nuclei fuse together to produce larger ones. Every element has its own particular type of nucleus.

The electric charge on a nucleus determines the chemical element produced, and the number of protons the nucleus contains determines this charge. The nucleus of a hydrogen atom is simply a single proton, and you might think that adding these together in sufficient quantities could readily build up any other nucleus. It is not that simple, because all other nuclei contain neutrons as well as protons, and hydrogen does not. As stars are mostly formed from hydrogen and there are no neutrons in hydrogen, then in some way the star must generate them."

The calm attentive figure of Paul, who had not been far from Aurora's side since she first encountered the Big Bang, took a step forward. He turned back several pages on his clipboard.

"You will remember how, in the period of the Big Bang before the formation of helium, neutrons decayed to protons until eventually the remaining neutrons were captured into nuclei, mostly helium. It was the weak interaction that did this. If protons are to convert back to neutrons, then circumstances must somehow be different than they were in the early stages of the Universe. Circumstances *are* different: The protons are now inside a star. In free space neutrons could decay to protons and not the other way round. This is because the neutrons are the more massive and when they change they release some small part of their mass energy. Small it may be but it is enough to provide the mass of the electron created, with a little over to drive the decay process.[11] The decay of a proton in free space would not provide any mass energy, quite the reverse in fact, but nuclei are produced in collisions in the hot dense center of a star, and protons *can* find the energy to decay to neutrons when they are inside a nucleus. This is because the energy of bound nucleons is reduced by some binding energy and the restrictions imposed by the Pauli Exclusion Principle may allow a lower energy state for a neutron than is available for a proton. If that is the case then a proton in a state of higher energy may decay to a neutron in a lower energy state.

"If some protons have been changed to neutrons, then other protons may combine with them to create helium, with two protons and two neutrons bound together. It is possible, but it is slow," he finished.

• • • • • • • • •

11 Weak nuclear decay has already been discussed in "Snow White and the Particularly Little People."

"Protons do not combine easily," carried on an impressive figure with a squared-off beard, a laurel wreath, and a toga who was grasping a handful of jagged thunderbolts. "They all carry positive charges and so repel one another. They may only interact if they get very close to one another, and this may only happen if they are moving fast enough to get close together despite the repulsion. If hydrogen gas is to fuse, it must be hot, so that its component atoms collide with great energy. If there

are to be many interactions of the protons, they must collide often, so the gas must be dense. With the weak interaction being an essential part of the process, it will be slow, so the gas must remain hot for a long time. None of this is a problem for a star," he added, looking disapprovingly at the protostar. He raised one of his thunderbolts and flung it arcing forward. There was a flash and the star began to shine with the intensity of a newly ignited nuclear furnace.[12]

"This is a relatively big star," continued a rather Eastern-looking deity with a pointed headdress and a generous quota of arms. "It was born when the galactic supply of hydrogen was dense and plentiful, and so there was plenty of material from which such a large star could be built. Officially it would still be classed as a dwarf, as are all the stars in the Main Sequence[13], but it is a big dwarf, some hundred times as large as your Sun. More to the point, it radiates a good million times more intensely. Consequently, whereas your Sun will shine for some ten billion years, such a great star will expire in a few million. This is still, you might say, a long time, but it is short enough for many such stars to have lived and died before your Sun was born.

· · · · · · · · ·

12 The nuclear ignition of a protostar does not actually require a thunderbolt from a demigod with attitude. When gravitational contraction of the core makes it dense enough and hot enough, then fusion becomes possible and just begins.

13 The Main Sequence of stars is mentioned in "Ali Gori and the Cave of Night" earlier in this book.

"Such a star will create many elements. At first it will burn hydrogen to give helium," he announced, ticking this new element off on one of his fingers. "It will create neutrons as required and in this process it will much enrich the small amount of helium that was initially present.

"When the star has finished burning hydrogen in its core, then it may begin to burn helium to form other elements, or rather their nuclei." He ticked each off on a new finger as he mentioned it. "Next come carbon and oxygen, the prime elements needed for your sort of life." He folded down more fingers " Neon and magnesium, silicon and sulfur." He listed all the elements up to the final and most strongly bound element of Iron. It was a good thing he was so well equipped with fingers.

"At iron the fusion process must finish, even in such a massive star whose gravity may create enormous temperature and density in its core. In iron the nucleons are as tightly bound as in any nucleus and any further accumulation of nucleons would release no energy. Fusion processes end at this stage and the central radiation from the core fails. The outer envelopes of the star still contain successive shells of fusion spreading out from the earlier core processes. The star becomes seriously unstable and..."

The hawk-headed divinity stood up with arms spread like wings and opened his beak "Ka – ," he screeched piercingly if rather unintelligibly. "Boom!" he added. As if on a signal the swollen star was blown apart in a monstrous supernova explosion that blasted some of the star's store of new elements into the surrounding space.

Now that the dramatic moment had passed, the varying divinities busied themselves with their next task. This turned out to brewing another pot of tea while discussing among themselves which star they should go and monitor next. Behind them glowed the expanding shroud of the great supernova.

As there did not seem to be much further information coming from that quarter, the girl and her companion moved away. "That supernova demonstrated the method of seeding space," remarked Paul. "If we are to deliver the sort of Universe that some of our clients want, then we must have a variety of elements present in it. If planets are to be born and support life, then they must condense from a cloud of gas and dust that already contains these elements. It is from such supernova that the elements are provided. Live people need dead stars."

Around them in that springtime of the galaxy, many such great stars were born, lived their brief brilliant lives, and spewed their contribution of elements into the surrounding medium. With every million years that flickered by, more of these stellar fireflies blazed up and then were gone; each one added its contribution to the mix of matter in space. Other stars formed as well, smaller, fainter, relatively inconspicuous but more persisting. The very smallest ignited and shone almost apologetically with a small light that nevertheless persisted on and on throughout the birth and death of many brighter stars. Stars of all sizes formed from the gas available, most of small and medium size now that the richest pickings from the primeval cloud had been creamed off and then locked up in the dying cores of those giant stars spawned in the dawn of the galaxy. Stars of moderate size, too, reached the end of their lives after many of their brighter cousins had ceased to be. These also delivered their smaller cargo of nuclei to the galactic mix.

The galaxy they occupied had by now taken the form of a great spoked wheel. There was the wide thin disk that spread on either side of the central bulge and carried the angular momentum of the condensing gas cloud. Now it was largely depleted of gas, being instead full of the stars that had formed from its condensation. Curved spiral spokes radiated from the central bulge. They were the peaks of compression waves that rotated around the galaxy and formed regions of denser gas in which stars were now selectively being formed. Paul drew Aurora's attention to a point about a third of the way out from the center of the galactic disk.

"It is now some ten billion years after the Big Bang," he remarked. "Doesn't time fly when you get involved with something. Here is a star forming now that may be of significance to you."

"Is it in some way remarkable, then?" asked Aurora.

"No, not really, it is a very average star. I just though you might have a certain interest in this one." He would say no more and they watched in silence as gas and dust condensed under gravity to form the increasingly hot and glowing blob of a protostar. There was a slight rotation in the gas from which the star was forming. In any rotating system there is an apparent outward centrifugal force[14] so that, just as was the case for the galaxy as a whole, it was difficult for all the material to condense down to a small volume. In the process of jostling and mixing within the gas, most of it managed to shed its rotation and condense down toward the center, leaving the bulk of the angular momentum to be carried by a tiny fraction of the mass. This fraction settled into a flat rotating disk surrounding the star. The overall form was broadly that taken by the entire galaxy during its creation, since the reason for the shape was the same. The disk was in due course destined to give birth to the star's quota of planets.

• •

Angular Momentum and the Solar System

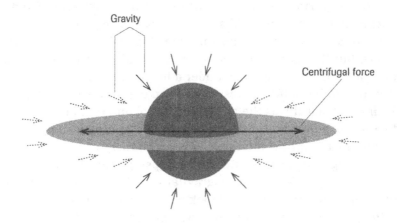

Stars form when galactic gas and dust condense into some central concentration. Gravity seeks to pull all the material down toward a central point, but this is resisted by the centrifugal force caused by the angular momentum remaining in the rotating gas cloud. Collisions within the cloud transfer much of this angular momentum to a small portion of the gas that forms a disk around the bulk of the star, while the remainder of the star is now able to collapse to a spherical ball. Centrifugal force keeps the disk material some way out from the center, but does not prevent it from collapsing under gravity to give a thin disk. The planets of our present Solar System are prevented from falling into the Sun by their angular momentum.

• •

• • • • • • • • •

14 Some people maintain that there is no such thing as centrifugal force. If you look at a rotating gas system from a non-rotating frame of reference then you just see gas molecules trying to escape along a straight-line path. In a frame that rotates with the system, however, there is apparently outward (centrifugal) acceleration and force to cause it. This is sometimes called a 'fictitious' force, but the NASA Internet site argues the force is real enough in a rotating system. Jet turbine blades do not tend to shatter for fictitious reasons.

"You say this is going to be an average star. Do you mean that it will be larger or that it will be smaller than my Sun at home?" asked the princess, who liked to get things straight.

"Oh, it will be very much the same. In fact it will be exactly the same, because this *is* your own Sun in the process of formation."

In the swirling disk the material was beginning to clump. Within the chaotic motion of the gas, some of the small percentage of atoms that were neither hydrogen nor helium were clumping together to form dust. Some portions of this dust moved more quickly and in different orbits than others and so there were collisions. In a gas this would normally lead merely to mixing of the atoms, but the more substantial dust coagulated slowly to form dust balls, much as happens under a bed in the absence of spring cleaning. At least, the process is slow on the timescale of beds. In the life of a star it was the mere blinking of an eye. The clumps of dust in their turn collided and coagulated to create small rocky pebbles that in turn combined to give small rocks.

Throughout these events the collapsing material of the star was becoming ever hotter as gravitational energy was released, and it glowed ever brighter. It took no active part in the formation of the planets other than providing light by which Aurora could watch the process, as well as a *solar wind* that blew the light gas away from the inner regions so that the careering stones could move more freely and rapidly between collisions.

Aurora looked at this chaos of hurtling boulders and noted, rather to her surprise, that on one of them sat a motherly looking woman busily engaged in her knitting. The girl and her companion drifted closer, following the circling path of the boulder until they were able to accost the lady.

"Who are you?" asked the princess. It seemed as good a conversational opening as any.

"Why my dear, I am Gaia. Or you may prefer to call me Mother Earth. I am the tutelary spirit of your planet." She absently lifted her feet out of the way of a plunging rock that shot past her.

"But there is no Earth!"

"Not yet perhaps, but that is all right. It is just a matter of time. It is quite common for mothers to have to wait some time before their children are born. I keep busy," she added, holding up her knitting. It was coming along quite nicely, growing longer by the moment. It took the form of a pair of intertwined helixes joined by many cross-links, all knitted from the complex fuzzy forms of molecules. Aurora had no idea what its purpose could be.

The protoSun was still shrinking and glowing ever brighter as its temperature rose. Though hotter, it was giving less radiation because it had less surface area to emit it, and the blazing rocks in orbit were actually becoming cooler. Eventually the gaseous core became hot enough for nuclear processes to ignite, and ignite they did. A new source of energy switched on in the core and the new star had a brief period of intense radiation before it settled down. This outward blast blew much of the dust and gas clear of the inner planetary region. The star soon reached a steady state where the pressure of the outgoing radiation nicely countered the inward force of gravity, and it settled into its long untroubled life as a nuclear furnace.

In the disk around it the boulders were still colliding and building up ever-larger masses. Some of the masses had become so large that their gravitational attraction was drawing in rocks, and so these masses grew ever faster.

Now there were only a few objects, but each of such size that you might well call them planetary masses. Where before the boulders had been irregular, all craggy angles and protruding lumps, now gravity was beginning to decide their shapes. Gravity felt that the proper shape for a blob of matter was a sphere, since this allowed as much of it as possible to get close to the center of mass. The forces between the atoms in the objects tried to keep them in the shape they happened to have when they were formed, but as their masses grew the planets deformed. Gravity overcame any resistance and forced them to be roughly spherical. There were a few irregularities, mountains and the like. To an observer on the surface the mountains would seem large enough, but they could not be too large because they were limited to the size that could be supported against the overwhelming pull of gravity. In the outer reaches of the solar system, in a volume that had not been so efficiently swept clear of gas by the initial blast of radiation from the Sun, planets also began to form. As there was mostly gas present the planets were essentially balls of gas and as there was so much gas they were large. One of them, Jupiter, was close to becoming a very small star in its own right.

Large rocks and asteroids still kept striking the surface of each inner planet, making the rock run molten at the point of impact and leaving deep craters in its surface. (The outer gas giants do not really have a surface.) In time this bombardment slackened and almost ceased, leaving planets that looked much the worse for wear. Most planets had an atmosphere, a mantle of gas that surrounded then and gave rise to some sort of weather. One planet also had a coating of water that formed a deep, wide ocean. On the scale of a galaxy or even of the solar system, this water was just a drop in the ocean, if you would pardon the metaphor, but seen from the surface it was pretty impressive. The actions of air and water gradually managed between them to smooth out the signs of impact on the planet's surface. The planet still did not look particularly attractive, but less cold and forbidding than its Moon, which, dry and airless as it was, retained the visible craters from its early bombardment.

"That is the Earth," Paul told the girl. "There you see a cradle in which life may grow, as promised in our prospectus," he added. "Yes," cut in Gaia. "There you will find that life develops. Come and see for yourself." She stretched out her hand and together she and the princess descended gently into the planet's atmosphere.

It was not a very nice atmosphere and not one that Aurora could have breathed with any comfort, or indeed at all. Nonetheless, for Mother Earth this was home and in her company even the volcanic eruptions and the unbreathable atmosphere did not seem too unfriendly. Somehow, when in her company, the girl was able to breathe and to look around her in what even seemed like serenity. What she saw was not encouraging: a barren and desolate landscape, lightning flashing overhead and the horizon aflame with active volcanoes. The bombardment by boulders from space had largely ceased, but that was the only comfort.

"I should think it will be a long time before any life develops here," Aurora remarked in a thoughtful voice, looking at the discouraging landscape around her.

"And there you are wrong," returned Mother Earth. "Life is already here. Look more closely."

She really did mean closely. The princess looked around and there seemed nothing animate, but somehow she was made aware of what she could not normally have seen, even with a magnifying lens. Around her in the rocks were organisms. Simple, undeveloped but alive in that they moved and reproduced and processed energy from their environment. Single-cell organisms that fitted into their surroundings as best they could, living in a world deficient in oxygen, in some cases even in the sulfurous waters near volcanic vents.

They were primitive but as time passed they developed. Initially it took them a very long time before there was much obvious change. Billions of years passed, a period that was significant even in the long lifetime of the Universe itself. Yet life still existed only as single-celled forms. They had changed and adapted dramatically, but it was hardly obvious to the casual observer.

Then suddenly, almost overnight on the cosmic scale of time, life proliferated dramatically. In mere millions of years a rich and diverse assembly of different creatures appeared in the Precambrian eruption. They presented a teeming variety of little beings. Some were not too successful, and after but a short time, they died out. Others remained and developed. Plants appeared and set about turning the carbon dioxide of Earth's atmosphere into the oxygen mix more familiar to Aurora. For some creatures, to which oxygen was a poison, this was bad news, but the change in the atmosphere suited the development of large oxygen-breathing animals who based their metabolism on this active gas.

Rich ecosystems developed and thrived until some disaster cut down the dominant species and others evolved in their place to suit the changing circumstances. There were several such extinction events. The last cut short the reign of the dinosaurs, huge creatures that ruled and fought without serious contention in their day, until a great meteor crashed into the Earth and made the day seem as night with the dust clouds it flung into the air.

The dinosaurs died but small scurrying mammals survived and began their own path to dominance. The dinosaurs had had size and power and ferocity. Some of the earlier mammals tried this path, but the prize went to a species noted mostly for inquisitiveness, for curiosity coupled with agile fingers. They developed tools, they developed languages, they developed cures for many illnesses, and they flourished mightily. Soon they had spread like a cancer across the entire Earth and ripped it apart in the process. Aurora and Paul watched their progress in the company of Mother Earth, who kept tut-tutting under her breath.

It was not all bad. These creatures, these humans, developed minds with intelligence, they developed art and religion and science, and throughout their history they had a mania for story telling. The three spectators narrowed their focus to one example, a small group who clustered around a Storyteller. They listened.

"And that is the story of Waking Beauty. It began long ago in a chaotic explosion of energy and for us, at least, it has led to the world we see around us. We know there is a vast universe around us and no doubt there is wonder aplenty among the stars. Is the tale of Waking Beauty now finished? Who can say? I can not, but it would be surprising if it were...."

INTRODUCTION TO

the sixth tale

"The story of the Universe has been a long one, and will be longer still," continued the Storyteller. "The Universe began small and has expanded for billions of years. It is far from ended and will expand for many billions of years to come. Perhaps it will expand forever. It all depends."

"Depends on what?" asked Joseph.

"It depends on gravity, as does so much in the Universe. The galaxies are all rushing apart as space expands. Gravity is working to slow down this dispersion, as every galaxy tugs on the gravitational bonds it has with its companions, and attempts to rein in their mad flight. In this the galaxies may or may not succeed. If there is enough mass in the Universe, and the pull of gravity is sufficiently strong, then the growth of the Universe will slow down and eventually reverse, with the galaxies then heading back toward one another. This is like the case of a ball that you have thrown high in the air, but gravity will eventually halt its rise and bring it falling back to Earth.

"If there is not enough matter in the Universe, then the bonds of gravity will be too weak, and the expansion of the Universe will continue forever. This is more like the case of a rocket fired upward with more than the Earth's escape velocity. Gravity could not hold it, and it would leave the Earth completely.

"If the Universe is to have such an indefinite future of ever-expanding space, you might ask, what is to become of the stars? At present, the Sun warms us during the day, and at night we can still see the stars shining brightly in the sky. It may not always be so. Prepare yourselves now to hear the forbidding story of 'Cinderenda and the Death of Stars.'"

CINDERENDA

and the

DEATH *of* STARS

(the life and fate of stars)

Cinderenda had gone to the ball. As a culmination of a series of remarkable events, too complex and indeed implausible to recount, she had finally made it. It was the night of her dreams and she danced and danced as if she would never stop. The evening passed all too quickly, however, and the hands of the clock moved inexorably toward midnight. At last the clock struck: one, two, three ... Twelve sonorous notes rang out and died away, their last echoes lingering in the air. As she heard the dying echo of the last stroke of midnight the girl noticed that the ballroom had become strangely empty. The sound had faded just to that point where she could no longer say whether the sound was still there or not when, without warning, the tall double doors at the end of the hall crashed open.

In the doorway was framed a tall figure in black robes that had the ragged texture of scorched cloth. Within his dark hood was the blackness of deep shadow, but as he turned his head, enough light penetrated to reveal a naked skull, a skull with the flaky consistency of fire-calcinated bone. The startled girl was gazing upon a personification of HEAT DEATH.

"Are you the Grim Reaper?" quavered the nervous girl.

"No, I am Heat Death, the Cool Stir-
rer," came the reply, in a voice so soft as
to be barely distinguishable from the level
of background sound even in that quiet
room. He gestured with the great wooden
spoon that he carried in place of the traditional scythe. "In me there is neither hot
nor cold, only a vast tedious tepidity. My
role is to mix up all extremes to give an
everlasting average." He turned his bony
face toward her, in the process revealing
two empty sockets, deep within which
could be seen the last dimming cinder of
a fiery spark.

"I am invincible, for I am Entropy: the
measure of disorder and of chaos. I pre-
side over the loss of any feature or variety
in the Universe. I am the nemesis of information. My Law is final because my Law is the Second Law of
Thermodynamics. Heat always flows from hot to cold, and in the
process cold becomes a little warmer and hot is not longer quite so
hot. Both have joined in a great uniformity of temperature. In the
process a certain variety, a certain contrast is lost to the Universe. I
am the sink of energy and there is no going back. Once it has
departed, energy does not flow back from the cold to restore the
hot," he added.

"How does that work?" asked Cinderenda, feeling rather confused but nevertheless glad that the apparition did not apparently
mark her immediate demise.

"Well," began her skeletal companion, "consider that bowl of
water by the fire." There were large fireplaces at either side of the
hall, and in each a fire had been roaring throughout the evening.
The black-clad figure stalked to one of these, pausing on the way
to pick up a bony handful of ice cubes from an ice bucket that had
been set out to chill drinks. He lifted the bowl and dropped the ice
cubes into it. There was no real surprise in the fact that they began
to melt. "It is all a question of molecules in motion. The water is a
collection of molecules and its temperature is given by the average
energy that its molecules possess. Observe!"

A hazy picture sprang up in the middle of the hall, like a
mirage or hallucination. It showed a teeming mass of little blobs,
molecules evidently. On the whole they moved around randomly,

at different speeds and in different directions, but also on the whole quite smartly. To one side there was a collection of them that stayed close to one another and oscillated to-and-fro. Cinderenda realized that these must be the molecules in the solid ice. The freely moving blobs collided persistently with those on the outside of the packed mass and in most cases knocked them from their position so that they joined in the activity. Soon all of the molecules had been freed and soon they were all in ceaseless random motion. On average the free molecules now moved a little more slowly, but it was not perceptible. They still collided with one another, pretty much as often as before. Sometimes a faster would strike a slower and it moved faster. Sometimes two that had been moving quickly would collide in such a way that one was left almost stationary. The girl could see that there was, however, an incredible number of the molecules, and as a consequence very many collisions, so that overall the effect averaged out, and on the whole there was no perceptible change.

"The mixing process is irreversible. With individual collisions the initial and final conditions might well be interchanged without seeming in any way unreasonable. In a collision between two molecules it is quite likely that one will end up with more energy and one with less; the tendency to average out is not evident for single collisions. It is just that it is unlikely that all of the molecules in any given region of significant size will end up with less energy than they had, and consequently those in another region will have more. Such correlated behavior is not impossible, but it is improbable. By improbable I mean that it would be vanishingly unlikely for such a thing to have happened on any of the planets in the visible Universe at any time in its entire past history. As I said, improbable," he remarked rather dryly. (Mind you, having no lips or tongue but only bare bone, there was no other way he could remark.)

"That a separation of more and less energetic molecules is possible, at least in principle, is illustrated by my associate, Maxwell's Demon."[1] He swept aside the hem of his black robe to

. . . .

1 This was a notion proposed, not entirely seriously, by the Scottish physicist James Clerk Maxwell. He imagined a tiny demon that could open and close a hole in a partition between two containers, allowing slow molecules to pass in one direction and fast in the opposite direction. The demon would thus reverse entropy, but in doing so he would be pretty busy.

reveal a small and bedraggled-looking demon carrying a sort of round shield and looking pretty unhappy. "He will now demonstrate what would be needed to reduce entropy and increase the order and contrast in a gas or liquid," he said through rather visibly gritted teeth. To Heat Death even the thought of entropy reduction was anathema. He picked up the small demon and held him over the bowl of water. The little sprite seemed to shrink in upon himself, and dwindled even further as the bony fingers released their grip dropped him into the bowl.

Within the vision of the hurtling molecules in the center of the hall, the tiny figure of the demon appeared, bearing his shield in front of him. In the huge assembly of colliding molecules in a gas or liquid, some are always slower and some faster, and he darted to-and-fro amongst them, holding his shield in the path of one particle after another. Where a slow molecule was moving toward the right he held his shield so that it rebounded toward the left. Where a fast molecule was moving to the left he intercepted it so that it was deflected toward the right. There was an inconceivably large number of molecules present even in the small sample that they viewed, but somehow the frantic little demon managed to intercept a significant fraction of them, and he systematically herded the faster molecules to one side and the slower to the other.

"There you can see the result of the demon's activity," said the tall grim figure. He indicated the bowl into which he had dropped the ice, and the girl gasped at the strange sight. At one side of the bowl a crackling rim of ice was forming, while directly opposite the water was gently steaming.

"This is not a sight that you see very often. In fact it is a sight that you never see. This is not because it is impossible, but there are so many molecules involved that it is not likely for there ever to be such a gross departure from the average condition. You saw how persistently our little demon had to impose an unnatural order on so many molecules, even though no energy was involved as the rebounds always kept their previous energy and merely changed direction."

"What do you mean, 'no energy involved'?" complained a weak voice. The bedraggled figure of the little demon had appeared in the ballroom and collapsed panting on the floor. "It is all very well for you to say 'no energy involved.' *You* didn't have to rush around after all those molecules. They aren't easy to catch, I can tell you."

The grim representation of Heat Death ignored him completely and went on talking to Cinderenda. "It is unlikely that the energy that has been distributed among particles roaming freely throughout the liquid may be again concentrated in one original source. In that fire," he said abruptly, pointing a bony digit toward the hearth at one end of the room, "the energy that has warmed the air in the room will not again return to the charred wood. When the fire has gone out, the ashes will never be reconstituted to make a fresh log of wood."

"But," began Cinderenda, who tended to be rather argumentative at times, even in the presence of daunting personifications of Universal Principles. "But you could sprinkle the ashes on the ground and a new tree could grow, you know."

"Yes, the presence of living organisms can give an increase in order, but not overall. The universal flow of energy to a condition of increasing disorder may be locally diverted into a side loop, but the overall entropy of the Universe is increased even here. Living organisms require that energy must flow through them. They must have food, a concentrated source of energy that they can burn, and they pass this energy on to their surroundings as waste heat. Overall, the chaos in the distribution of heat is increased in the process. The Second Law of Thermodynamics applies to a closed system, one without energy coming in or going out all the time. If you were isolated in a container without any input of food or air, you too would enter a state of increasing entropy. It is known technically as being dead."

· ·

Entropy

Entropy is difficult to describe. There are mathematical formulae, but they are far from intuitive.

It is usually described as a measure of disorder. If you have ice and water in separate places, you have rather more order than you do after they mix.

Increasing entropy has also been described as the loss of information. To say that you have ice in one place and warm water in another gives you more information than to say that you just have slightly cooler water.

· ·

Cinderenda rather wished she hadn't raised the subject. The discussion seemed to be taking a direction too much in keeping with her companion's appearance. For the last few minutes they had been standing by the fireplace in which the fire, previously crackling merrily, was going out. Now the last visible sparks died at the ends of blackened branches, and there was little left but ash.

"That is the fate of any fire, however large," commented the grim figure by her side. "It is the fate of any region where energy is more concentrated. It will lose its distinction, and a great uniformity will be achieved. That is the Second Law of Thermodynamics and there is no escape. Sooner or later any conflagration will exhaust its fuel and will die. That is true for the fires in this hall that have provided you with heat and light during the night. It is equally true of the fire that provided you with heat and light during the day."

He made a gesture with a dark-garbed arm and the drapes across the tall windows flew back to reveal the night sky beyond. "I mean of course the Sun. There you see other suns, the suns of distant planets other than yours. People speak of renewable energy sources, energy extracted from wind and waves and directly from the sunlight that falls upon the Earth. They call this renewable energy on the assumption that the Sun and its radiation will always be there. It won't. The Sun will continue to burn for a long time, some five billions of years, but in due course it will exhaust its fuel and, like any other fire, it will die. Come and see the future of the great lanterns in the sky."

To Cinderenda's considerable unease he reached out a bony hand and gripped her firmly by the arm. Together they rose from the floor and wafted toward a tall window, passing through the glass pane as if it were unreal, or perhaps they were. Up they floated into the night sky. Velocity and indeed the passage of time itself were impossible to estimate. Did their journey take an eternity or did it take no time at all? She was not sure, but sooner or later, she found herself, together with her companion, floating unharmed in space some little way from a blazing star. This also had a companion star some way off, but less vivid than the one nearby.

"That is a star much like your Sun. This one happens to have a binary companion, circling around it in a mutual orbit. but that does not much affect its behavior. It is currently well through the initial hydrogen burning stage in its path of nucleosynthesis."

"Nucleosynthesis, what is that?" asked Cinderenda naturally enough.

"Perhaps I may answer that," broke in another voice. Such an interruption was not perhaps what one might expect in empty space, but there floating beside them was a figure in flowing robes decorated with a variety of exotic symbols.

• •

Nucleosynthesis

The Big Bang left the Universe full of hydrogen, quite a lot of helium, and a smattering of other light elements. Useful materials such as oxygen and carbon were made later in the hearts of stars. A star is a lot of gas, mostly hydrogen, which initially condenses onto some form of clump. As the gas collapses inward, the gravitational energy released makes it hotter and hotter, and of course it also gets denser. Eventually it becomes sufficiently hot and dense that collisions in the gas may cause nuclear reactions and thus build heavier nuclei.

A collision between two protons may very occasionally lead to the formation of a deuterium nucleus, a deuteron. This consists of one proton and one neutron. This can happen if, during the collision, the weak interaction should convert one proton to a neutron. Once neutrons have been created it is comparatively easy for two deuterons to combine and form a helium nucleus, with two protons and two neutrons. From this, other nuclei may be built in stages if the star is large enough and hot enough.

• •

"I am a Stellar Alchemist," he announced. "Nucleosynthesis is my specialty. That is the name that describes the building of new elements, the production of heavier nuclei from the Universe's initial stock of hydrogen. As you know, chemical behavior depends on the number of electrons in an atom, and the number of electrons depends on the electric charge of the atomic nucleus.[2] As goes the nucleus, so goes the atom.

"In order to have the sundry materials that you find on Earth, you must have a wide selection of nuclei with which to build the atoms. In particular, if you are to get the amazing variety of compounds that are needed for organic life, you require carbon and oxygen and a variety of other elements such as nitrogen, calcium and phosphorus, as well as hydrogen. Hydrogen is already provided in the early stages of the Universe, but the other elements must be created in stars. The Alchemists of old tried to convert lead to gold, but the stellar alchemy is much more wonderful. This is for two reasons. Firstly, it creates the very elements of life from the hydrogen gas that pervades the Universe. Secondly, it actually works.

"Now," he continued in the determined tone of someone who has rather a lot to say, "there is an obvious problem with the task of building up heavier nuclei from hydrogen. Every single type of nucleus but one contains neutrons. Unfortunately hydrogen is the one that does not. The Universe in its early stages is singularly devoid of neutrons, so it is necessary somehow to produce them. Enter the weak interaction!" He paused dramatically for an audience reaction. The best the girl could manage was an expression of patient interest.

"The weak interaction can change protons to neutrons and neutrons to protons. Unfortunately for our objective, what the weak interaction has been doing throughout most of the history of the Universe has been to change neutrons into protons. This is simply because neutrons are slightly more massive than protons. When a neutron converts to a proton, you also get an electron to balance the electric charge of the proton. The neutron is sufficiently heavy that it can provide the rest mass of the electron and still have a bit left over. Mass is energy, you know, and the surplus will go into the kinetic energy of the particles, the confusion of

• • • • • • • • •

2 See "Snow White and the Very Small People," earlier in this book.

movement that provides an entropy increase and encourages the decay of neutrons.

"Inside a star, things are rather different from free space." Abruptly he reached into the center of the star and withdrew a handful of–something. This was an unexpected and surprising feat. Cinderenda tried very hard to see how he did it, but her eyes were quite unable to follow his actions.

"Basically, the center of a star is hot," he continued. "This is a first requirement if protons are to combine, as the protons all carry the same positive electric charge and so repel one another. If they are to get close enough together for the strong and weak interactions to operate, they must collide with enough energy to overcome this repulsion. In the hot center of a star some few of the protons will have enough energy."

He opened his hand to reveal a cloud of particles, presumably protons. It became clear that he had not simply removed them from the star's center, as they would rapidly have dispersed into space. It appeared that he had in some way made visible to her a small portion of the interior of the star. The particles behaved as if they were still in the dense core, squashed in upon one another and engaged in the frantic dance of high energy. Some would leave the visible region and others enter to take their place as they moved in and out of the sampled region within the nuclear furnace. On occasion the protons would approach one another. When they did, they slowed as they got near because their positive charges pushed them apart. Almost always they stopped and turned back, but some were moving so fast that even with this loss of speed, they came so close that they effectively touched before they were forced apart. They had come close enough for the strong interaction to bind them, but as long as they remained protons it failed to do so because there are no bound states that contain only protons, and so they separated again. Neutrons are necessary to dilute the repulsion of the protons electric charges.

• •

Proton decay inside nuclei in stars

In free space neutrons may decay to protons. A neutron is sufficiently heavier than a proton that it can release sufficient mass energy to create the masses of the resultant proton and electron, with a bit left over for kinetic energy of the decay products. Nature loves to release kinetic energy, as it can provide a huge number of indistinguishable final states of motion and so increase the entropy of the system.

In order for a lighter proton to decay to a heavier neutron, something must reverse the balance of energy. Somehow the creation of a neutron must release kinetic energy, and this can happen if the neutron is much more tightly bound within a nucleus than a proton can be.

Within nuclei, the Pauli Exclusion Principle limits the states available to protons and to neutrons, since protons and neutrons form sets of identical particles. Protons are not identical to neutrons, however, so there are two separate sets restrictions.

Within stars nuclei are formed in high-energy collisions in the hot, dense cores. In the potential nucleus that is briefly formed during a collision, the lowest energy state for a proton may already be filled, but an equivalent neutron state may be empty. If the proton converts to a neutron by a weak interaction, then kinetic energy may be released and a stable nucleus created.

Such weak interactions are infrequent, and the whole process is slow, but stars are very patient things.

* *

"The weak interaction is able to convert protons to neutrons inside a star.[3] However, the weak interaction is weak and so the decay is slow. It is so slow that a million collisions may not produce a single conversion from proton to neutron. Eventually, though, the change will happen, and when it does a proton and a neutron bind together as a deuteron, the nucleus of deuterium—heavy hydrogen. As this bound state is created, energy is released to heat the gas. A deuteron may collect a further proton to create a nucleus of helium-3, a nucleus that contains two protons and one neutron. Further collisions may result in the creation of a helium nucleus with two protons and two neutrons. In this last process, there has been no need to create new neutrons as deuterons already contained protons and neutrons in equal numbers, as do many heavier nuclei. The helium nucleus that remains at the end of this process is powerfully bound, and much binding energy has been liberated with its formation, serving to heat the stellar core.

"The production of helium nuclei is as far as the process goes at this stage. As the concentration of helium increases, some of the

* * * * * * * * * *

3 This was discussed in the previous tale "Waking Beauty" and at some length in the box entitled "Proton decay..." just above.

new helium nuclei may come together within the range of the nuclear force and even form a compound nucleus for an instant, but as it happens no nuclei containing eight nucleons–some mixture of eight neutrons or protons–is stable. Any newly formed nucleus will soon fall apart, and the chain of nuclear buildup snaps at this link. It might seem to be but random chance that there are no stable nuclei with eight nucleons, but on this chance depends much of the nature of stars throughout the Universe.

"However, the nuclear furnace continues to burn and convert hydrogen to helium. Since it relies on the weak interaction to create neutrons, the furnace burns slowly, but it contains so much hydrogen fuel that even this faltering fire can release much energy that is eventually radiated as sunlight. The core of a star of moderate size, such as the Sun, burns so slowly that it may burn for ten billion years and allow time for life to develop on any planets it possesses."

"Even so," broke in the somber figure of Heat Death, who had been standing, dark and silent, to one side during the Alchemist's lecture. "The star cannot burn forever. It may take ten billion years, but eventually, as with any bonfire, the fuel in its core will be exhausted and the fire will die. There will be hydrogen remaining outside the core that will continue to burn, and a shell of fusion will spread outward, but the hydrogen fusion in the core will cease. With any normal fire this would mean that the region would cool down, but stars are a bit different. When the fire in the core dies out, the core will shrink and consequently get hotter. It is a funny thing with gravity," he mused, temporarily losing the thread of his discourse. "I always seem to have difficulties with it. On the really grand scale, the laws of thermodynamics are not so obvious.

"In a local isolated sample, one that is not in a condition of change and has settled down into a state of equilibrium, my laws work well. Entropy increases inexorably and the law of conservation of energy–the First Law of thermodynamics–is scrupulously obeyed. A fast moving object like a cannon ball will, when it strikes a castle wall, release its energy as sound and heat, plus the kinetic energy of stone fragments that it knocks out. This last involves replacing the energy that bound the stones within the castle wall. For the rest, the sound and the heat, the energy has gone into the movement of atoms within the air and in the wall itself. The situation is complicated, but the balance works out meticulously. The total energy after the collision is just as it was before.

"On the scale of the Universe, things are different. Nothing is truly local, as the reach of gravity is enormous and it involves such things as the expansion of space. When the Universe became transparent after the Big Bang it was flooded with light, light of enormous energy, light that contained by far the greatest concentration of energy in the Universe. This light had for the most part no further interaction with the matter in the Universe. As time passed, space expanded, the wavelength of the light expanded with it, and the energy of the photons that comprised it went down in proportion. After more than ten billion years this vast blaze of radiation is virtually undetectable to you, being nothing but a pervasive microwave background. Certainly the energy is now spread out over a vastly larger space, but it is not just that. The original photons have not interacted and each has much lower energy than it started out with. The energy of all these photons has not been conserved. It has actually decreased.

"On an isolated local scale, though, the laws of thermodynamics work well enough. A star system is small enough and sufficiently isolated. Certainly it is bathed in the light from other stars and in the universal background radiation that I have just mentioned, but within a star these are negligible compared with the radiation pouring out from its core. Also, the star is enfolded in the gravity of the Universe that affects the expansion and the shape of space, but locally this is as nothing compared with the gravity of the star itself. To a very good approximation the star may be considered to be in total isolation from the rest of the Universe, and the laws of thermodynamics will apply. Its central fire, as with more mundane fires upon Earth, will burn out.

"Throughout its long period of apparent stability, the material of the star has been supported against the steady pull of gravity by the fierce outward blaze of radiation. When this falters, the core resumes its former collapse. Its fall releases gravitational energy and as the core shrinks, it heats. As long as no new process comes into play, there is no limit on this gravitational collapse. The material at the center of the star becomes ever hotter and more dense."

"Eventually," interrupted the Alchemist, "the nuclei at the heart of the star will be packed so closely and moving with so much energy that when two helium nuclei collide and combine there is a good chance that yet another nucleus will collide with them before the combination has time to fall apart. This second collision has increased the size of the newly formed nucleus and in so doing brings it within the range of stable nuclei. The helium fusion in the star has created a nucleus of carbon, with six protons and six neutrons. It may readily go on to generate oxygen, with two more each of protons and neutrons. You should find that particularly interesting," he remarked, turning toward Cinderenda.

"Why should I?" she asked.

"Because to a fair approximation, they *are* you. Carbon and oxygen, together with the universally prevalent element of hydrogen, form the main components of organic life. Physically, at least, you are your atoms.

"As I said, helium fusion cannot happen unless the collisions are frequent. It will thus be a fast process. Beginning with a 'helium flash,' the star will give out more heat, and because of the extra radiation from the fast process in the center as well as the shell of burning hydrogen moving outward, there will be more pressure on the envelope of the star."

"What effect does that have?" asked Cinderenda. It sounded like quite a dramatic event. This was confirmed by a new voice that broke in, though it did not seem to be addressing her.

"Good evening ladies and gentlemen. This is your correspondent Roger Rover, reporting to you from the death of a star. It was not a great star. It was quite an average, ordinary star, in fact; but now it is dying."

The girl looked around and saw, balanced on a small planetoid nearby, a keen-looking young man who was clutching a microphone and speaking earnestly into it. A cable trailed through the void to a nearby spaceship, in front of which a space-suited gang were operating some sort of camera. The reporter did not appear to be wearing any protective clothing, but a faintly glowing envelope could just be seen surrounding his figure.

"After billions of years of shining steadily as a yellow dwarf, casting its benign light upon the planets that surround it, now it has run amok. Runaway fusion of the helium store that it has painstakingly built up over its long life is causing it to expand into a red giant. Its size will balloon until it has enveloped the innermost of its family of planets. Let us hope that there was no life on them, ladies and gentlemen, because surely there will be none now."

As he spoke the star swelled enormously, expanding out to envelope the innermost of its family of planets, just as he had said. It was now huge compared to its original size, and its light output had greatly increased. This enormous output was spread across the huge area of its surface, which because of its greatly increased size was actually cooler than that of the normal dwarf star that it had been. It now glowed distinctly red. The main sequence dwarf star had become a red giant.

"I am reminded of a verse that I read once," he went on reminiscently.

> *What does this furious star portend?*
> *Sure it gives a lovely light.*
> *But oh, my foe and oh, my friend,*
> *It will not last for more than a few million years.*

He turned away from the camera and clicked off a switch on his microphone. "Are you sure about that?" he asked his crew. "It doesn't sound quite right to me. Well, never mind, no one will be listening too closely anyway." He turned his microphone on again.

"The star now burns at such a furious pace," he went on, "that it will exhaust itself that much more quickly, in but a few million years." He switched off his microphone again. "We can edit out the next bit before we transmit the program.

We will just cut to the final core collapse. OK folks, let's have a tea break." The film crew all retired inside the space ship. Cinderenda was fascinated, as she had never witnessed a million year tea break before.

Time passed and somehow, under the influence of her immortal companions, it did not seem to take so very long. Soon enough the roving reporter was back atop his rocky planetoid.

"It is getting very tense now, ladies and gentlemen. The helium fusion also is dying out in the core, though helium and even hydrogen still burn in shells closer to the surface of the star. It's any millennium now folks. Yes, there it goes!"

In some way, due to the wonders of an advanced technology, the camera managed to focus on the burnt-out core within the still blazing outer shells of gas. The core began to shrink and did so rapidly, heating and glowing brighter as it did. In the process it emitted a blast of energy that caused the star's outer shell to shred off into space. The star had suffered such collapses before and they had continued until the center reached such a level of heat and density that some new nuclear fusion processes could begin. This time, before the temperature could reach quite high enough, the collapse halted! No new process of burning began in the star's centre. Nothing happened this time to give a new source of radiation whose outward pressure could support the collapsing mass of the core. There was no such support, but still the collapse halted.

"What has happened?" the girl asked her companions as she realized that the situation was in some way novel. It was neither of her companions but the reporter who gave her the answer as he addressed his unseen audience.

"As you can see, the star has halted in its collapse and indeed will now shrink no further. This time it is not because of the outward pressure of radiation, as no new fusion processes have begun. It is because of something that you would not expect to affect a star. The final state of a star, which is a very large thing, has been determined by something that comes from the realm of the very small. The collapse has been halted by quantum effects, by the physics of atoms."

• •

White Dwarfs and the Exclusion Principle

It is remarkable that the final state of something as large as a star may be governed by quantum theory, the physics of the atom. This however is the case.

Quantum theory says that in a limited space the states available for particles are distinct and generally separated in energy. In atoms the energy separations are typically a few eV^4. In nuclei, which are much more compact, the energy separations are typically millions of eV, written as MeV.

A volume of astronomical size will still give discrete quantum states, but they will be *very* close together. Usually you would be quite unaware of them. The occupation of these states by electrons is restricted by the Exclusion Principle, just as in an atom.

It would take an extraordinarily large number of electrons to fill all the states up to ones that had a significant amount of energy. A star *does* contain an extraordinarily large number of electrons, and when a star collapses to a white dwarf the lowest quantum states are all filled.

White Dwarfs and the Uncertainty Relation

When a star collapses to become a white dwarf, it fills all the lowest energy levels that correspond to the volume it occupies. If the star is still

• • • • • • • • •

4 An eV is a measure of energy often used in atomic physics. It is the energy one electron gains when it falls through an electrical potential of one volt. As eV units are used for the energy of single electrons, they are pretty small. One calorie is equal to twenty six billion billion eV. Anything expressed in eV is going to look pretty fattening.

to collapse, then the quantum states themselves must become more compact, as there are no lower levels left for electrons to occupy.

If a particle falls under gravity toward some mass, it releases energy of an amount inversely proportional to its distance from the center of that mass. In a white dwarf, both electrons and nuclei release energy as they fall. When the quantum state in the star contracts, the momentum of the electrons is inversely proportional to the size of the mass. An electron of a given momentum will have a higher velocity than would a heavier particle, and so it has a higher kinetic energy. Further contraction of the electron quantum states would require its kinetic energy to rise more quickly than energy would be released from gravitational binding energy. As the energy required is not available, no further contraction may happen. The white dwarf cannot collapse any further.

A normal star, such as our Sun, is in a state of unstable equilibrium. Its collapse is temporarily prevented by the outward pressure of radiation from the nuclear furnace in its core.[5] A white dwarf, on the other hand, is inherently stable. It is in effect a very large atom, with gravity taking the place of the electrical interaction in the atom, and there is no obvious reason why it should not last forever.

• •

He paused for dramatic effect, and Cinderenda had time to note that he was reading his detailed information from an enormous prompt screen that towered in front of his space ship.

"The stellar core has now been squeezed down so far that the electrons are filling all the lowest quantum states available in that central region. Yes, even something as large as the central region of a star will have permissible quantum states that may be occupied by electrons. There is an enormous number of such states, but then there is an enormous number of electrons to occupy them. Electrons obey the Pauli Exclusion Principle." He pronounced these names with some pride as he read them out. "All of the lowest energy states have been filled with electrons; there can be no further contraction unless the states themselves contract. More localized states have higher momentum." Again he made this announcement with a hint of triumph. "The kinetic energy would

• • • • • • • •

5 OK, it will last for billions of years. On the grand scale of things that is still temporary.

rise more quickly than energy would be released by gravity, and so the star will not shrink. It cannot.

"This star has reached its final state. The core is now composed of *degenerate matter* and cannot collapse further. It has become a White Dwarf. This is rather like an atom in that the Heisenberg Uncertainty Principle decides its size. The star's center has become a sort of atom, a very large atom certainly, but that is because gravity is so much weaker than electromagnetism that you need a lot of matter for gravity to produce the necessary compression. This White Dwarf is a stable object and should last indefinitely, slowly dimming as it cools. It will steadily lose the heat that remains from its period of collapse and will eventually become a dark star. Because the star is so small, it loses heat very slowly. It is estimated that no White Dwarf has yet cooled to invisibility since the beginning of the Universe," he announced with the ready confidence of someone who has his facts in a script in front of him.

"And so, Ladies and Gentlemen, we may now depart and leave this poor remnant in decent obscurity. We can expect no further action–or can we?" He paused in his narration with a dramatic flourish: "Remember, Dear Audience, that this star had a companion. It was one of two stars circling around one another. The White Dwarf and its more conventional companion waltz sedately, bound as partners by the tether of gravity. As the White Dwarf orbits, it raises tides on the fiery surface of its companion, much as the circling Moon raises tides upon the Earth's oceans. As these tides circle around, the two stars lose energy and their orbits slowly spiral closer. Then, with little warning, the second star enters a helium burning stage and swells up to a red giant, as stars are wont to do. As its outer envelope shreds off, the attendant White Dwarf captures some of it."

As he spoke, Cinderenda could see the other star flare up, and streams of gas spiraled from it onto the White Dwarf. This blast of new material tumbled down into the deep gravitational well of the White Dwarf and spread over its surface, adding to its mass. As the White Dwarf grew ever heavier, it showed signs of disturbance.

"What is happening now?" asked Cinderenda. Once again the reporter took up his commentary.

"The star is changing before your eyes. As its mass increases, so does the gravity that is attempting to crush it. Electrons are forced into states of ever-greater energy and eventually they become highly relativistic. Why does that affect the collapse of the star you might ask? Simply that it removes the very reason why the

White Dwarf could not collapse before. When the star shrinks the electron states become more compact and their momentum rises. If the electrons are relativistic then their energy rises at much the same rate as their momentum, and gravity *is* now able to provide this energy. The star will suddenly collapse."

· ·

Neutron star formation

A white dwarf star is stable, at least as long as the kinetic energy of electrons would have to rise at a greater rate than their momentum, since this is proportional to the rate at which collapse under gravity could release energy. This is so as long as Newtonian mechanics effectively applies—as long as the electron velocities are not too high.

In a stable white dwarf the electrons in the highest energy quantum states still have energies less than their rest mass energies. For larger stars, more of the electrons are forced into states of high energy, and at some stage they will have energies that will greatly exceed their rest mass energies. They will no longer obey Newtonian mechanics—they will be *highly relativistic,* and their kinetic energy will rise at much the same rate as their momentum. Gravity *can* provide energy at this rate, and the star will collapse dramatically.

The next stage will be determined by nucleons, not by the now ultra-relativistic electrons. The star will still be mostly hydrogen, so the nucleons will be dominantly protons. The electrons in the white dwarf prevented the protons from collapsing because the relatively powerful electrical interaction prevents the positive and negative charges from becoming too widely separated.

Collapse is complicated by the presence of both positive and negative charges. If there were only massive neutral particles, they could collapse until they were supported by the same sort of quantum effects as control a white dwarf, but on a much smaller sale, because massive particles can have high momentum at relatively low kinetic energy. If the protons and electrons were to combine to form neutrons, a great deal of energy would be released from the gravitational collapse, and this is enough to provide for the excess rest mass energy of the neutrons.

When a white dwarf collapses, electrons and protons of its hydrogen mostly convert to neutrons, and it shrinks to a relatively tiny size. Where the white dwarf is a sort of atom bound by gravity, the neutron star is a sort of nucleus.

Should a neutron star be too heavy, then even the massive neutrons become highly relativistic and the neutron star will in turn collapse. Now nothing can halt the collapse, and it becomes a black hole.

• •

With shocking abruptness the star did just that. From a bright nearby disk it shrank quickly to a tiny dot. This was not very easy for any observer to follow, as such a rapid collapse of an object of stellar size liberated huge amounts of gravitational energy. Much of this was given off as radiation in various forms and in amounts of which there is no conceiving. The star had become a Supernova, a beacon that would light up a galaxy and could be seen from other galaxies. The flashbulb light died away to leave an expanding cloud of gas that would persist for millennia, a great ring growing steadily and illuminated initially by the heat left from its formation.

As the incredible blaze died away, Cinderenda–who after all had never really been there–found that she had been in no way harmed by it. She looked toward the place where the reporter had been and saw a mirror-surfaced ovoid. This was clearly a protective shield of some sort and it gradually became transparent to reveal a somewhat

ruffled young man. His face was blackened and he was frantically
beating out small flames that flickered in his hair.

"Well, that was exciting," he announced rather wanly. "That
was a Type I supernova, folks. Remember that you saw it first on
this channel. This is Roger Rover returning you to the studio." He
switched off his microphone in some relief, looked around, and
noted that the cable from his microphone stopped abruptly a few
meters away. Of the space ship to which it had originally been con-
nected there was no sign. "Hey!" he exclaimed, "My [*expletive
deleted*] crew has gone off and left me!"

Feeling reasonably confident that his employers would not
lightly abandon such an obviously expensive talent and would
soon come back for him, the girl felt she could safely put him out
of her mind, and she turned her attention back to her two companions. "What has become of the White Dwarf?" she asked. "It has
quite vanished."

"Oh it is still there," replied the Alchemist. "It is just more difficult to see. In fact it is still a stable ball of degenerate matter like a
White Dwarf, but not quite the same *sort* of matter. It has become a
neutron star."

"What is that?" the girl asked predictably.

"Have patience and I shall tell you," rejoined the Alchemist,
also quite predictably. "You have been told how the White Dwarf
could no longer support itself against gravity when its electrons
became relativistic, and so will shrink. As it continues to shrink so,
the density and also the gravitational binding energy of the particles will increase. Any attractive interaction that binds particles, be
it gravity or the strong interaction, will generate a binding energy.
Now, you should remember what happens in a star where neutrons may be created." The Alchemist opened his hand to reveal
again the selection of closely packed energetic particles that he
had displayed before, when he was talking about the buildup of
heavier elements in stars.

"In free space neutrons decay to protons and not the other way
around because neutrons are more massive and can release
energy to drive their decay. In the heart of a star, collisions of protons can lead to the formation of nuclei in which the binding
energy of neutrons may more than compensate for their greater
mass and then protons may decay to create them. I told you of this
if you remember," he said, staring at her rather fiercely.

"Within this terminally collapsing star the particle binding
energies produced by *gravity* have now become so great that an

electron and a proton may combine through the weak interaction to give a neutron. The more massive neutron has a greater gravitational binding energy and this will drive the process. The center of the new star converts to tightly packed neutrons and it becomes a neutron star. Although the momentum of its particle states is higher than that for the electrons in a white dwarf, the mass of the neutrons is so much greater than that of electrons that their motion is now non-relativistic.

"The quantum effects that prevented the collapse of a stable white dwarf apply again and so the neutron star also is stable. Where the white dwarf is a sort of star-sized atom the neutron star is a sort of enormous nucleus. In the same way that the nucleus of an atom is so much smaller than the atom itself, a neutron star is tiny in comparison to most stars. A neutron star more massive than the Sun would be a mere ten kilometers in diameter. It is of course a great deal larger than an atomic nucleus, but you need an awful lot of neutrons for gravity to give the same sort of energies as does the strong nuclear interaction."

Together the small group drifted in toward the center of the devastated star as the Alchemist continued his discourse. The tall somber figure of Heat Death had said nothing for some time, but it was still hard to be unaware of his dark presence. Whenever Cinderenda glanced at him she felt tepid shivers running up and down her spine.

The neutron star was now visible as a tiny dot. Then, as they moved around it, suddenly it flashed briefly and brilliantly.

"What was that?" exclaimed the girl, momentarily dazzled. As she spoke the bright flash came again and again, like a speeded-up lighthouse marking the shore on which the star had foundered.

"Why that was the neutron star, of course," replied the Alchemist. It is rotating very rapidly, you know. Stars generally rotate, but they are big and it takes them a long time to go around. Very large things generally change their appearance rather slowly, since no part of them may move faster than the speed of light. If a star shrinks down to become a neutron star, it will keep its angular momentum, so as the size decreases, the rate of rotation will increase. Where a normal star will take many days to rotate once on its axis, a neutron star may spin many times a second."

"But why did it flash like that?" interrupted Cinderenda.

"I am coming to that," answered the Alchemist rather crossly. "A star will usually have a magnetic field, and if the star shrinks down to a neutron star, the field is compressed along with it and becomes very intense. Fast-moving electrons will spiral around in the magnetic field, and in the process emit *synchrotron radiation.* This will be emitted in a narrow beam along the direction of the magnetic field. This beam sweeps round and round as the star spins, like the beam from a manic lighthouse. When this happens, the neutron star may be seen from a great way off as a *pulsar.* When a star much larger than the Sun reaches its end, then often its core is too massive to remain as a white dwarf and then it may collapse to a neutron star," concluded the Alchemist, somewhat breathlessly.

"Is that how all massive stars end up?" asked Cinderenda, as something seemed to be required of her.

"No, not the most massive." This reply came in the calm, dull voice that they had not heard for some time. The personification of Heat Death had re-entered the conversation.

"For the most massive of stars, the end is even stranger. If you look at the sky, some of the stars that you normally see are not normal stars. They are the brilliant extroverts of the heavens, stars some hundred times as massive as your Sun and able to give off a million times as much light. They are quite rare, if only because they use themselves up so quickly with their extravagant brilliance. Such stars move through the same sequence of nuclear burning and nuclear creation as their lesser siblings, though they go further and create all nuclei up to iron, which is the most stable of all.

"In general, gravity is a problem to me, as I said before," confessed the grim representation of thermodynamics, pausing in his life history of a giant star. "The Second Law applies to isolated equilibrium systems with no energy flowing in or out. When gravity comes into the picture, it is never clear when a system is isolated, and in an expanding Universe you do not reach an equilibrium—energy changes with time. In a restricted region, however, the Second Law is a useful approximation and stars do burn out. Huge stars are plentifully supplied with fuel at their beginning, but they, too, burn out like any smaller star. More quickly, in fact, because of their spendthrift brilliance. In their final stages they will explode as supernovae and in the process pile nuclei together more rapidly than the compound nuclei can decay, and so leap across unstable nuclei to form long-lived heavy elements up to uranium.

"In their brief dramatic lives these great stars create the nuclei of every element you would desire. In their fiery ending they blast much of this material out into the wider galaxy to form the basis for future planets and any rocks, seas, and even life thereon. The rest of their substance they leave behind them as a cinder, as do smaller stars. This residue is now itself massive, many times as heavy as your Sun. It is much too heavy to make a stable white dwarf, and it is too heavy even to survive as a neutron star. Just as a white dwarf that is too heavy will collapse, so a neutron star that is too heavy will in its turn be unstable and continue to contract. You might ask what now can prevent it from collapsing indefinitely." Ever obliging, Cinderenda did ask just that.

"Why, nothing. There is nothing to prevent it from shrinking indefinitely. In fact, nothing *could* prevent it. When gravity becomes sufficiently strong, then no repulsive force may counter it. Indeed, any repulsion would only hasten the collapse, since contracting against a repulsive force stores energy, energy is mass, and further mass can only speed the collapse. The star is doomed to go on collapsing. Will its fall ever come to and end? The answer seems to be no. The star will shrink and shrink, and theory predicts that it will shrink right down to a dimensionless point of infinite density. It is destined to become a singularity, a point of disjunction in space-time. Such a thing is anathema to the physical world, but fortunately it is not a naked singularity. Its abhorrent final state is hidden from us by a veil of gravity. It carries with it all that it has made and fades from the sight of men and of the Universe itself.

"You may know," he went on more soberly "that an object on the surface of a planet, or in the gravitational field of any other mass, is bound by gravity with some amount of energy. If you were to throw that object so hard that it would leave the grip of gravity, then you would need to give it a kinetic energy greater than the binding energy of gravity. As a consequence, although it will slow as it rises, it will still have enough energy to escape with some velocity left over. That is to say, you must give the object the necessary *escape velocity*. In the case of this star collapsing without limit, the escape velocity will become greater than the velocity of light, and the star will become a *black hole*. No light, or for that matter anything else, can escape from it. It has in effect isolated itself from the rest of the Universe.

"Nothing that gets too close to a black hole can escape. At some distance from the center of the hole, known as the *event horizon*, gravity has become so strong that from within this horizon not even light may leave. Around the hole, SpaceTime is twisted and distorted. From a point just outside the event horizon light may escape to the outside, but only within a narrow window that points straight out. Light that leaves in any other direction will curve round and fall into the waiting black hole, which will absorb it forever. Light that passes close to the hole will curve around it and, if it passes close enough will fall into it and be captured. At a distance of twice the radius of the event horizon the light may circle round the black hole like a satellite about a planet. This is the closest orbit that anything may make, and solid objects would have to station themselves yet further out, since they must travel more slowly. "[6]

They had drifted in toward the black hole that was now clearly visible in front of them, or rather not visible. It could be perceived as a disk of nothingness outlined by distant stars. Close to the disk the stars appeared to crowd together and streaks of light showed where starlight had been warped in the intense gravitational field of the mass.

The figure of Heat Death turned and looked intently at Cinderenda from his empty eye sockets, themselves black holes in his smooth white skull. "It is not simple to speak of distances and times in the vicinity of a black hole. The circumference of the event horizon of a star-born black hole may be fairly small, while by some reckoning the distance to its center is very large. A distorted SpaceTime does not sit well with the geometry of the old Greek Euclid. It is not just distance but times that are elusive. Time here is much in the eye of the beholder, though no illusion. Were you, rather unwisely, to launch yourself directly into a black hole, you would soon enough reach its center.

This would be a region where gravity would vary so enormously with the distance from the center that your feet would be pulled with much more force than your head. You would be torn to shreds upon the instant, though no report of your fate could reach the outside world, for you would be inside the event horizon."

"In such a situation is there no possibility whatever that I might escape?" asked the Cinderenda. "Would there be no hope at all of reprieve?"

"I fear not," replied her companion somberly. "For the very structure of SpaceTime would conspire against you. The Universe can be merciless and implacable. You see here Nature Dread in Truth and Law.

. . . .

6   This was the situation for the spaceship described in "Jack and the Starstalk."

"However," he continued, "should a companion be watching your suicidal dive from a point safely away from the hole, he or she would see your descent slow down as you approached the event horizon; you would never reach it, but everlastingly inch toward it. At the position of the event horizon, the time dilation, as seen by someone at a considerable remove, would

become infinite, and time would not appear to pass at all, as seen from the outside. Your observing companion would not actually *see* this of course, as the frequency of any light coming from you would also decrease to such a degree as to be invisible. This slowing applies to time as it appears to your remote observer; for you, your dive into the unimaginable center of the black hole would pass all too quickly."

The figure of Heat Death, who had been close by the girl's side for so long, abruptly launched itself away from her, sailing in toward the small black blank in the star-filled view before them. With his long dark robes trailing away behind him, the white skeletal figure seemed to slow as it approached its destination. For a while it appeared to hang, tiny and almost motionless in space, before it darkened and faded from sight.

"There you see a demonstration of the time dilation effect," remarked a flat emotionless voice from nearby. Cinderenda turned her head to see the familiar figure of Heat Death floating nearby.

"What ... how?" she stammered. "How can you be here? I have just seen you fall into that black hole from which you tell me nothing can escape."[7]

"True enough, but you must realize that I am not a thing or even a person. I am a characterization of a universal principle. I am omnipresent. I am everywhere and so I am still here by your side, wherever else I may be.

. . . .

7   A suggested method by which energy may escape from a black hole requires that particles should be both inside and outside the event horizon. Quantum states are not well localized. Pairs of virtual particles (discussed in the tale of Snow White and at greater length in my book *The Wizard of Quarks*) may be spread across the event horizon. In a steeply changing gravitational field, by the process of *Hawking radiation,* one particle may gain enough energy to escape while its companion falls into the hole.

"Am I in the black hole?" he continued. His cowl had fallen back from his head and he nodded his bare skull in the direction of the singularity. "It is difficult to say what is inside a black hole since anything inside it is lost forever. No information, no hint remains on the outside of the nature of the objects that fell into it to provide its mass. It is said that a black hole has no hair, as it is so devoid of external features." The girl looked at the smooth, bare top of her companion's skull and wondered if he was also a personification of a black hole.

"As viewed from the outside, and that is the only way you can view it, a black hole will reveal only its total mass. The size of the event horizon is a measure of this. If the matter were originally rotating, then the black hole will have a distortion from perfect symmetry that reveals this. A black hole might in principle have an electric field, though that is unlikely. That is all however. All detailed information about its constituents has been irretrievably lost. This may be seen as an ultimate triumph for entropy. As material falls into a black hole, all information about the material's nature and properties will be lost. As it gains mass, the black hole will grow, and indeed its surface area is a measure of its entropy. Should black holes collide, they may join to create a bigger one and in the process their total surface area, and hence the overall entropy, will increase.

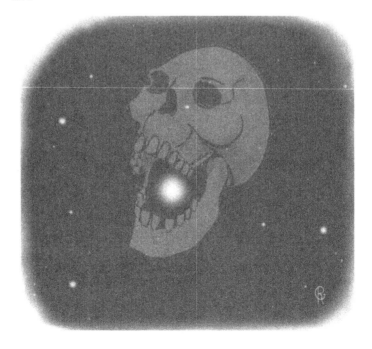

. . . .

Black holes and entropy

Entropy is a measure of featurelessness, of the loss of information. There is nothing more featureless than a black hole. It has mass and possibly angular momentum. It might theoretically carry an electric charge, but that's it. Anything that falls into it, however interesting or complex, is lost to the Universe at large.

Gas, stars, personal computers, or political candidates, if they should fall into a black hole, they are no longer knowable by the outside world. You cannot broadcast a manifesto from inside a black hole.

A black hole is the quintessence of entropy. It reveals nothing of what it absorbed. It has been shown theoretically that the surface area of a black hole's event horizon is a measure of its entropy.[8] The more a black hole absorbs, the more feature and information it takes from the Universe, the larger it becomes.

. . . .

"In the center of some galaxies there are gigantic black holes that have absorbed much of the material from which the galaxy was formed and even some stars that were present earlier in the galaxies' lifetime. A possibility for the future of the Universe is that more and more stars will end their life collapsing into singularities, and that these will gradually capture and absorb the other material remaining, so that at the end the whole vast Universe may contain nothing but these remnant black holes, with the old Universe locked away within them. Then entropy will have its ultimate triumph."

This apocalyptic comment was all the more chilling in that it was uttered in such a dull, tedious, and almost inaudible voice. Cinderenda looked at the stars around her and saw that behind one after another loomed the skull of Heat Death, gaping wide and swallowing whole each blazing orb. In every case, as the jaws of Heat Death closed around the star, it faded away to a dark nothingness. As she watched, she got the sensation that she was viewing a final closing down of creation, not with a bang but a whimper.[9]

. . . .

8 Don't ask me to explain how this comes about. Ask Steven Hawking. He thought of it.
9 OK, this phrase is not original, but it did seem to fit. (The prediction is far from certain, by the way.)

epilogue

"That's really sad!" declared Rachel, stunned by the conclusion of the last story. "Is that all the Universe has to look forward to, even if the end will not happen for a long time?"

"It may seem a bleak ending," conceded the Storyteller, smiling wanly at his audience. "To some people the Second Law of Thermodynamics does seem to suggest a doctrine of final decay—as the Universe approaches a state of equilibrium with no distinguishable features—much of it locked away inaccessibly in black holes—nowhere any possibility of rescue—a dreary and featureless eternity at the end of it all—a final state of extreme entropy—no hope ..." The Storyteller seemed rather listless and bewildered as this speech drifted along, but then he paused and, rather surprisingly, smiled again, this time quite brightly and cheerfully.

"But consider this. The Universe *began* after the Big Bang with a mix of particles in completely featureless and random equilibrium: a state of apparently *total* disorder. Over a long period of time, however, through the operation of gravity and the various other processes that have affected the physical world, the Universe has managed to ..."

He paused again and looked at the bright faces of his audience, and beyond to the tree-covered landscape, the distant hills and townships, and above them all to the twinkling stars in the sky. "It has managed to make itself *rather interesting!*"

further reading

I have written three other books in a similar vein to this one, namely *Alice in Quantumland, Scrooge's Cryptic Carol,* and *The Wizard of Quarks.* All of these are published by Copernicus Books, and can be found in many bookstores and libraries, or viewed at the publisher's website:

www.copernicusbooks.com

I also list here a few relatively recent books that I read and found valuable as I was writing the present work. The list is in no way a complete survey of the literature:

Jones, Barrie W. 1999. *Discovering the Solar System.* John Wiley & Sons.

Longair, Malcolm S. 1997. *Our Evolving Universe.* Cambridge University Press.

Begelman, Mitchell, and Martin Rees. 1998. *Gravity's Fatal Attraction: Black Holes in the Universe.* W.H. Freeman & Co.

Silk, Joseph. 1997. *A Short History of the Universe.* (Scientific American Library Paperback No. 53.) W.H. Freeman & Co.

Kaler, James B. 1998. *Stars.* (Scientific American Library Paperback No. 38.) W.H. Freeman & Co.

The very latest information on cosmology is best found on the Internet. Because sites change from time to time, my intention is to list a few current ones on my own website:

www.phy.bris.ac.uk/allegory.

Shortly after a book is published I tend to think of things that I wish I had included. I hope to include these afterthoughts, as well, on my website, together with any other material that I feel my readers may find helpful.